Birds, bogs and forestry
The peatlands of Caithness and Sutherland

David A Stroud, T. M. Reed,
M. W. Pienkowski and R. A. Lindsay

Edited by D. A. Ratcliffe and P. H. Oswald

LIVERPOOL INSTITUTE
OF HIGHER EDUCATION

Order No.
L 360

Accession No.
137900

Class No.
Q 910.5321

Control No.
44900.

Catal.
 8 NOV 1991

Contents

	Summary	5

1 Introduction
1.1	The peatlands of Caithness and Sutherland: the rise of conservation problems	11
1.2	Survey methods	16

2 The blanket bog
2.1	The Peatland Survey of Northern Scotland	17
2.2	The origin of the peatlands in relation to forest history	18
2.3	Peat formation and bog types	19
2.4	The international distribution of blanket bog	24
2.5	The significance of surface patterning	24
2.6	Threats to and losses of peatlands in the British Isles	28

3 NCC's Upland Bird Survey in Caithness and Sutherland
3.1	Introduction	31
3.2	Programme of ornithological surveys	32
3.3	Composition of the peatland breeding bird fauna	34
3.4	Waders	36
3.5	Other waterfowl	40
3.6	Raptors	44
3.7	Other species	46
3.8	Conclusions	50

4 Overall distribution and numbers of peatland birds in Caithness and Sutherland
4.1	Introduction	59
4.2	Wader densities and habitat characteristics	59
4.3	The identification and mapping of areas of peatland suitable for breeding waders	68
4.4	The use of landforms to estimate abundance and losses of waders on peatlands	72
4.5	The numbers of other peatland breeding birds	75
4.6	Summary of the ornithological interest	75

5 Unsurveyed habitats and species groups
5.1	Lochans, lochs and rivers	77
5.2	Invertebrates	77

6 Effects of afforestation on the ecosystem
6.1	Effects on birds	79
6.2	Losses of peatland birds	82
6.3	The lack of compensatory gain in forest birds	83
6.4	Effects on vegetation	85
6.5	Effects on abiotic features	89
6.6	An overall assessment	90

7 International implications
7.1	The 'Bern' Convention on the Conservation of European Wildlife and Natural Habitats	91
7.2	The 'Ramsar' Convention on Wetlands of International Importance especially as Waterfowl Habitat	92
7.3	EEC Directive on the Conservation of Wild Birds	93
7.4	The World Heritage Convention	94

8	**Birds and bogs: their conservation needs**	95
9	**Acknowledgements**	109
10	**References**	111
11	**Appendix** **Methods of ornithological surveys**	119

Summary

The blanket bogs of Caithness and Sutherland

The cool, wet and windy climate of northern Scotland has led to the development of extensive tracts of peatland which cover the landscape of most of Caithness and Sutherland. This is possibly the largest single expanse of blanket bog in the world and the largest single area of habitat in the United Kingdom that is of major importance on the world scale, because of its global scarcity. According to evidence from within the peat, the current treeless condition over most of the deep peat area is not due to historical clearance of natural forests by man. The peat bogs are a natural Post-glacial climax

Blanket bog is now under intense threat, mainly from afforestation. The area of the Caithness and Sutherland peatlands already lost to forestry represents perhaps the most massive single loss of important wildlife habitat since the Second World War

vegetation type (section 2.2). They represent the largest area of actively growing acid bog in Britain, and their vegetation is composed of plant communities which have no counterparts elsewhere, except in Ireland, where very significant losses have already occurred. The high degree of surface 'patterning' or pool formation on the flatter areas of peatland ('flows') is of particular conservation significance. The pools support a specialised range of mosses (especially species of *Sphagnum*) and vascular plants, and they provide essential feeding habitats for wetland birds.

These blanket bogs support a particularly varied northern type of bird fauna not found in identical composition elsewhere in the world. They hold important breeding populations of golden plover, dunlin, greenshank and arctic skua. The lochs and smaller dubh lochans support breeding red-throated and black-throated divers, greylag geese, wigeon, teal, common scoters and red-breasted mergansers. Rare breeding waders include Temminck's stint, ruff, wood sandpiper and red-necked phalarope. Raptors such as hen harrier, golden eagle, merlin, peregrine and short-eared owl also use the bogs as breeding or feeding areas (Chapter 3). Many of these species have their main distribution in subarctic and arctic areas, and the peatlands of Caithness and Sutherland have considerable ecological affinities with the arctic tundras (sections 2.5 and 3.3).

Considerable bird populations are present (section 4.2 and Table 0.1), although densities of individual species are often low. We have estimated that some 4,000 pairs of golden plover, 3,800 pairs of dunlin and 630 pairs of greenshank breed on these peatlands. They hold considerable proportions of the European Communities' breeding populations of several wader, waterfowl and raptor species, including 66% of greenshank, 35% of dunlin and 17% of golden plover (Table 8.1).

The United Kingdom has accepted international commitments for the conservation of wetland habitats and bird species (Chapter 7). Many of the breeding bird species on these peatlands are listed in Annex 1 of the European Economic Community's Directive on the Conservation of Wild Birds. Under this Directive Member States have entered into an obligation to take special steps to protect the habitat of such species and other migratory birds. The 'Bern' Convention requires conservation of threatened bird habitats, while the peatlands meet the agreed criteria of internationally important wetlands and hence qualify for designation under the 'Ramsar' Convention. The World Heritage Convention requires the conservation of natural features of outstanding universal value for the heritage of mankind.

Of all terrestrial habitats in Britain, these blanket bogs are the largest example of a primaeval ecosystem. They are of global significance, with both structural and biological features peculiar to this country. However, their continued survival is now under threat.

The threat

Coniferous afforestation is destroying these peatlands. Both the state forestry enterprise and a private sector company have obtained extensive land holdings on the open market and now own or manage extensive areas of Caithness and Sutherland. Whilst two decades ago threats to these peatlands on this scale would have seemed inconceivable, land-use change unprecedented in its speed and scope is now in progress (section 2.6). The total area of blanket bog in Caithness and Sutherland is estimated to have covered 401,375 ha before afforestation. Since then, at least 79,350 ha have been planted or are programmed for planting, about 67,000 on peat.

Successful afforestation requires deep-ploughing and draining, which disrupt water-tables and surface flow patterns and lead to longer-term erosion, shrinkage, deep cracking and oxidation of peat (Chapter 6). As the trees become established, higher evapo-transpiration rates lower the

water-table further and change the soil structure. Ground vegetation is eliminated when the forest closes into thicket after 10-15 years.

Afforestation is inimical to the survival of moorland breeding birds. Whilst some species may persist within young plantations for a short period, they disappear once the forest canopy closes into dense thicket. From this stage until the mainly unthinned forests are clear-felled, the transformed habitat excludes all the species of open moorland except the few which may be able to nest in trees, but even these will depend for feeding on the extent of open ground remaining unplanted. The peatland bird assemblage is replaced by one of woodland which, as well as being almost entirely different, is of much lower conservation value because most of its species are so widespread and common. Neither the vegetation nor the bird assemblage show more than insignificant recovery to the previous peatland types during subsequent clearance phases in the forest rotation.

Afforestation can cause non-breeding and territory desertion by wide-ranging predatory and scavenging birds, such as golden eagle and raven, even when only part of the open hunting range is planted. Moorland bird densities thin out over a perimeter zone beyond the forest edge, and suggestive indications that this is partly an effect of increased predation from crows and foxes are under investigation. Evidence for marked changes in vegetation on unplanted ground adjoining new plantations already exists and is consistent with a surface drying effect. Small enclaves of unplanted 'flow' and other habitats left unplanted within the forests are of doubtful value, especially in the longer term, either as examples of peatland vegetation and structure or for their bird assemblages.

The physical and chemical effects of coniferous plantations also extend far more widely than the plantations themselves. After ploughing and draining there are increased sediment loads in streams and lochs, faster run-off, and other alterations to the hydrology of catchments. During the early stages of plantation growth there can be short-term eutrophication owing to fertiliser application and run-off. This can affect nearby acid bogs by wind-drift and cause profound nutrient pollution of streams, with growth of algal blooms (section 6.5). All these factors reduce the quality of breeding habitat for water birds such as dippers and divers as well as affecting commercial freshwater fisheries, which are of major economic importance in Caithness and Sutherland (section 5.1).

The impact of afforestation of the blanket bogs in Caithness and Sutherland causes a serious net loss of nature conservation interest which may extend beyond the areas actually planted.

The surveys

The Nature Conservancy Council (NCC) has undertaken extensive survey work concerned with both the breeding birds and the vegetation of the peatlands. The methodology of NCC's Upland Bird Survey and Moorland Bird Study is presented in the Appendix. The Royal Society for the Protection of Birds (RSPB) has also conducted surveys of breeding birds, some of them using this methodology (section 3.1). Between 1979 and 1986 a sample of some 19% of moorland in Caithness and Sutherland was surveyed quantitatively for its breeding bird populations. The surveys were used to assess the nature conservation significance of the moorland bird populations throughout the two districts of Caithness and Sutherland.

The principal species of breeding waders — golden plover, dunlin and greenshank — all exhibit a considerable range of breeding densities, which can be related to variations in vegetation and its structure (Chapter 4). Associations between breeding waders and habitat features allow the interpolation of results over unsurveyed areas, using map evidence alone. This method has been tested, and the results used to obtain population estimates (Table 0.1). By examining map evidence from areas that have been afforested it is possible to estimate the numbers of the main wader species previously occupying them and thus the loss of populations through afforestation. There have been losses of up to 19% of golden plover, dunlin and greenshank (section 6.2). A disproportionate amount of prime habitat for waders has been ploughed and planted: foresters and these birds are in direct competition for the same areas of peat bog.

The results of NCC's Peatland Survey of Northern Scotland are briefly summarised (Chapter 2) and will be published in detail elsewhere (Lindsay et al. in prep.).

Conservation aims and the future

The future of these peatlands is uncertain since afforestation continues unabated. This report is not concerned with the questions that have been raised elsewhere concerning the economic and social justifications of such upland forestry (National Audit Office 1986), but rather seeks to present the conservation case for the protection of the peatlands (Chapter 8).

Previous conservation measures consisted of identifying exemplary sites to represent the range of interest in the peatland ecosystem, with the intention of conferring special protection on these as National Nature Reserves (NNRs) or Sites of Special Scientific Interest (SSSIs). Not only is this an extremely arbitrary approach in the particular situation, but reassessment in the light of fuller surveys and international evaluation has shown it to be quite inadequate in meeting the conservation need. The Caithness and Sutherland peatlands are now regarded as having a national and international importance which lies in their total extent, continuity and diversity as mire forms and vegetation complexes and in the total size and species composition of their bird populations.

Forestry interests regard all plantable land outside specially protected areas as potentially available for afforestation and have already planted up to the precise boundary of some SSSIs. Most of the remaining peatland area is plantable, and under present government policy and financial rules there is little to prevent the whole of this becoming afforested, outside the present limited area of SSSIs and NNRs. The nature conservation case is that the losses of habitat and birds already sustained on these internationally important peatlands are so heavy that any further afforestation is unjustifiable. These losses are compounded by the parallel losses to moorland habitats (including blanket bog) and birds which continue apace through further afforestation in other parts of Britain and abroad. Maintenance of nature conservation interest could be achieved simply by retaining the existing pattern of land-use. Such an approach is compatible with existing agricultural and sporting interests. Indeed, an integrated conservation policy for these areas could be of advantage to these other land-use interests.

NCC's surveys enable the extent of the nature conservation interest of these peat bogs to be quantified, but they have also revealed the rapid and continuing losses caused by afforestation. Already, habitat supporting nearly 19% of the three principal breeding waders has been destroyed or programmed for planting, and only eight of 41 hydrological systems in Caithness and eastern Sutherland have been left free from afforestation. The area lost to forestry — most of it since the passing of the Wildlife and Countryside Act 1981 — represents perhaps the most massive single loss of important wildlife habitat in Britain since the Second World War. Decisions to promote appropriate conservation measures are needed promptly if the losses already sustained are not to increase.

Table 0.1 Estimated proportions of national and European Communities' populations of selected bird species breeding in the Caithness and Sutherland peatlands

	Estimated Caithness and Sutherland breeding population (pairs)	Estimated British breeding (pairs)[1] population	Proportion of British population on Sutherland and Caithness peatlands	Proportion of European Communities' population on Caithness and Sutherland peatlands	Source[7]
Red-throated diver	150	1,000-1,200	14%	14%	Gomersall, Morton & Wynde (1984)
Black-throated diver*	30	150	20%	20%	Campbell & Talbot (1987)
Greylag goose	c. 300	600-800	43%	—[2]	Owen, Atkinson-Willes & Salmon (1986) Sharrock (1976)
Wigeon	80	300-500	20%	20%	Sharrock (1976)
Common scoter	30+	75-80	39%	16%	Thom (1986) RSPB and NCC (unpublished data)
Hen harrier*	30	600	5%	1%	Newton (1984)
Golden eagle*	30	510	6%	<1%[3]	Dennis et al. (1984) Watson, Langslow & Rae (1987)
Merlin*	30	600	5%	4%	RSPB (unpublished data) Bibby & Nattrass (1986)
Peregrine*	35[4]	730	5%	<1%[5]	Ratcliffe (1984)
Golden plover*	3,980	22,600	18%	17%	Piersma (1986); this report
Temminck's stint[6]	<10	<10	—	—	Rare Breeding Birds Panel (1986)
Dunlin	3,830	9,900	39%	35%	Piersma (1986); this report
Ruff*[6]	<10	10-12	—	—	Piersma (1986); this report
Greenshank	630	960	66%	66%	Piersma (1986); this report
Wood sandpiper*[6]	<10	1-12	—	—	Rare Breeding Birds Panel (1986)
Red-necked phalarope*[6]	<10	19-24	—	—	Rare Breeding Birds Panel (1986)
Arctic skua	60+	2,800+	2%	2%	Furness (1986)
Short-eared owl*	50	1,000+	5%	4%	Sharrock (1976)

* This indicates species listed on Annex 1 of the European Economic Community's Directive on the Conservation of Wild Birds (79/409) as requiring special protection measures, particularly as regards their habitat under Article 4(1). Other listed species are migratory and require similar habitat protection measures under Article 4(2).

1 This excludes the whole of Ireland.
2 EC population uncertain owing to unknown proportion of feral birds in other populations. The population in north-west Scotland is the only one thought to be natural, owing to separation from others.
3 Most of the EC population is of the south European race *homeyeri*; Britain holds all of the EC population of the nominate race, 6% of which occur on the Caithness and Sutherland peatlands.
4 The total Caithness and Sutherland population has increased to c. 60 pairs since the 1981 survey (Dennis pers. comm.), but no corresponding national total is available.
5 Most of the EC population consists of the Mediterranean race *brookei*; Caithness and Sutherland peatlands hold 5% of the EC population of the nominate race.
6 For reasons of confidentiality it is not possible to indicate precise numbers and distributions of these species breeding in Sutherland and Caithness. The Scottish populations of Temminck's stint and red-necked phalarope are the only representatives of these species breeding within the EC. Wood sandpipers breed in one other region of the EC but the Scottish population is important in EC terms.
7 Data for all species were also taken from Cramp & Simmons (1977, 1980, 1983, 1985).

Introduction

1.1 The peatlands of Caithness and Sutherland: the rise of conservation problems

Mountains and moorlands are the most extensive natural and semi-natural terrestrial habitats remaining in Britain, covering at least one quarter of the country (about 6-7,000,000 ha), mainly in the west and north. Their vegetation consists of grassland, dwarf shrub heath, peat bog and marsh, alpine 'meadow', moss and lichen heath, and fell-field. The local pattern in upland vegetation depends on topography, especially steepness and range of altitude, on the nature of the underlying rock and derived soils, and on land-use. Superimposed upon these local variations are broad geographical trends reflecting the main gradients of climate, of decreasing temperature from south to north, and of increasing oceanicity (particularly increasing rainfall, atmospheric humidity and windspeed) from east to west and towards coasts.

Blanket bog, the particular subject of this report, has developed naturally where cool, wet climatic conditions have favoured waterlogging of the ground and accumulation of plant remains as peat. It is a formation especially associated with flat or gently sloping ground, but occurs at increasingly low elevations and on increasing inclines as climate becomes more oceanic towards the north and west of mainland Scotland and its islands. Blanket bog covers the high plateau of Dartmoor, but in extreme oceanic areas such as Sutherland, Caithness, Lewis and Shetland and the west of Ireland it is extensive on low-lying moorland down almost to sea level. In these situations it represents a northern tundra-like ecosystem which has developed in these more southerly latitudes because of the highly Atlantic climate.

The largest expanse of blanket bog in Europe, and possibly the largest single area in the world (Figure 1.1), is where the low, rolling moorlands of east Sutherland descend gradually into the plains of Caithness. Further west in Sutherland, blanket bog is still

Figure 1.1 World distribution of blanket bog on Peters' projection, which shows correctly proportioned land areas.

The dark areas show those regions within which blanket bog occurs: the total extent of blanket bog is smaller. Note that blanket bog occurs almost exclusively between 40° and 60° latitude north and south on ocean seaboards

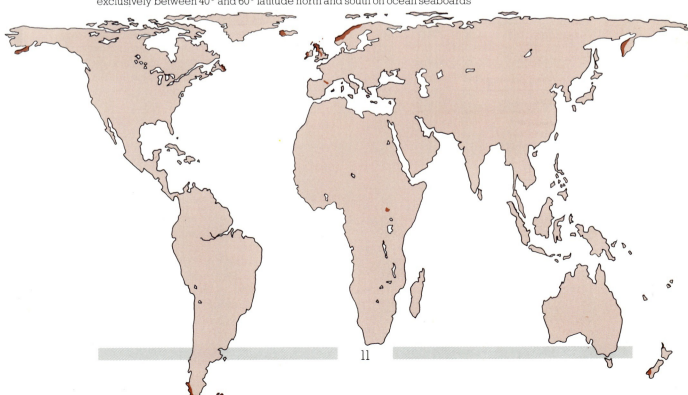

widespread but becomes more dissected by higher mountain ranges. The whole area of the Caithness and Sutherland blanket bogs, lying between 10 and 450m and originally covering 401,375 ha (Figure 1.2), is considered in this report. This area extends well beyond the largest single expanse of bog, sometimes called 'the flow country', occupying much of the area east of a line from Tongue to Lairg (Figure 1.3) or an even more restricted area (Royal Society for the Protection of Birds 1985).

'Flows' are flat or almost flat areas of deep bog which are especially extensive in this region and have in many places developed intricate surface patterns, in the form of complex pool systems. These patterned flows show wide variation in the size, shape and configuration of the pools and intervening ridges or hummocks, and they are of great scientific interest for their hydromorphology (Lindsay, Riggall & Burd 1985). Associated with this structural diversity is a distinctive and varied set of plant communities with dynamic successional relationships and composed of a flora which includes species of different biogeographical affinities. There are also numerous lochs, of widely varying size, and moorland stream and river systems. Within the accompanying fauna, the breeding birds are of outstanding interest: they represent a particularly varied 'tundra' type of assemblage and include nationally and internationally important populations of various species, as well as several national rarities.

The structural, vegetational and faunal variety are all closely interrelated, and survey information will be presented to identify and evaluate their interest. One of the most notable features is that the wet flow ground represents an unusually large area of natural habitat in this country, where so much of the land has been profoundly modified by past human activity. Although moor-burning has affected much of the total peatland area to some degree, quite large areas of the wettest ground have remained relatively undisturbed, and grazing by large herbivores has been light because of the naturally low carrying capacity. This is, indeed, a region which has largely escaped the more intensive modern kinds of land-use which have affected so many mountain and moorland areas elsewhere in Britain — upland farming and pasture improvement, water and hydro-electric supply, mineral extraction, military training, and the heavier kinds of recreational use. Its nature conservation interest has survived under a combination of traditional management for low-intensity sheep-farming, sporting interest in red deer, red grouse, salmon and trout, and local small-scale peat-cutting.

Concern over nature conservation on the Caithness and Sutherland peatlands has increased in parallel with the rapid advance of commercial afforestation in the region during the last decade. After a long period of forestry expansion elsewhere in the uplands, during which there was little interest in the planting of deep, wet bogs, a combination of silvicultural and technological advance has quite rapidly transformed the situation. Both state and private afforestation have spread rapidly, benefiting from a combination of advantageous grant-aid, tax concessions and land-market factors. A significant proportion of the peatland area is now owned by forestry interests, if not actually planted (Figure 1.4). Because much of the peatland area is at a low elevation, a large proportion of the total area is potentially plantable and therefore at risk. Afforestation causes a transformation of the peatland ecosystem and is regarded as almost totally destructive to its nature conservation interest (Nature Conservancy Council 1986). The replacement of these unique habitats by an extremely widespread and also artificial type of forest ecosystem is regarded by nature conservationists as a very substantial net loss of wildlife and environmental value.

The novelty, pace and scale of afforestation of the Caithness and Sutherland peatlands took conservation

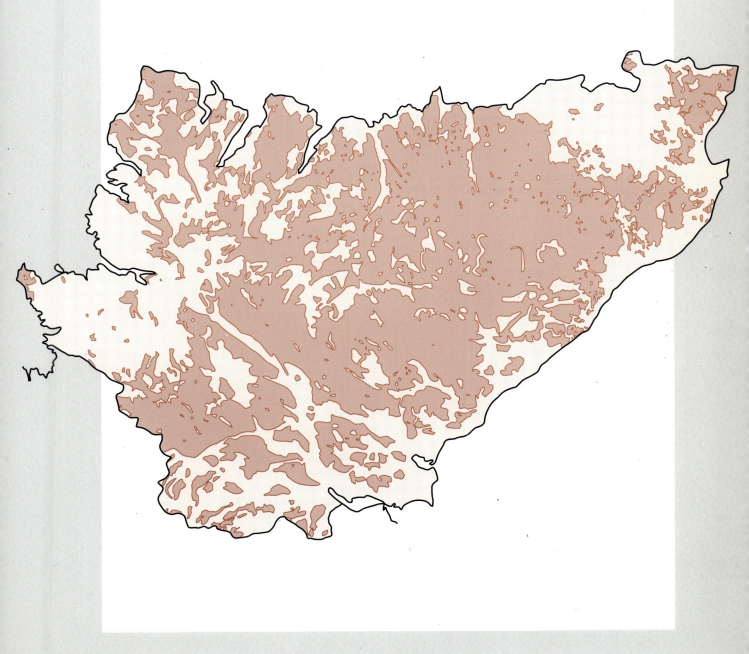

Figure 1.2 Distribution of blanket bog in Caithness and Sutherland before afforestation.

The map of peatland is derived from the soil categories of the Macaulay Institute for Soil Research. All of soil types 3, 4, 4d and 4e were included, with some combinations of other categories where the slope has allowed peat formation. From Lindsay *et al.* (in prep.)

Key

Blanket bog

Figure 1.3 Caithness and Sutherland, showing localities mentioned in the text

Key

--- Moine Thrust

Figure 1.4 Distribution of forestry in Caithness and Sutherland in relation to the blanket bog shown in Figure 1.2.

Areas shown are either in Forestry Commission ownership or have Forestry Grant Scheme approval or are dedicated private woodlands. Map as at January 1986, since when there have been many further FGS applications.

Key
- Blanket bog
- Plantations

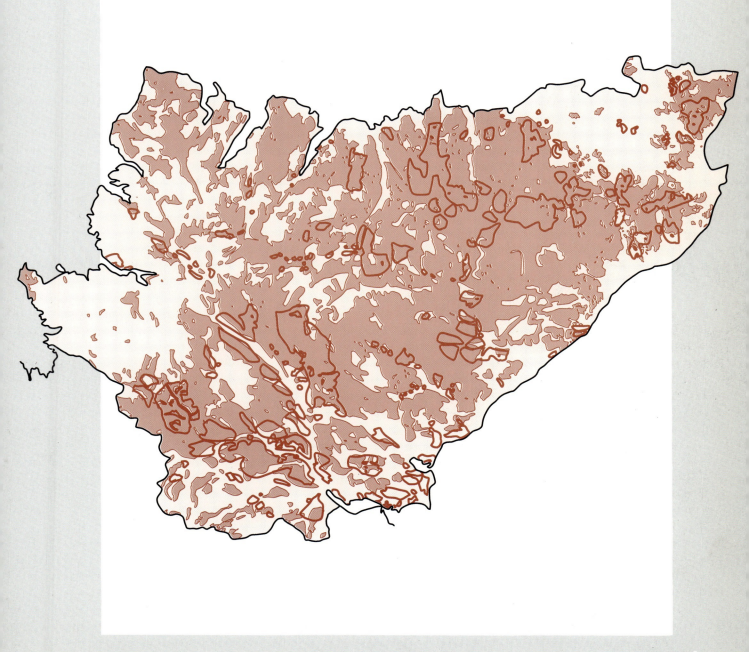

interests somewhat by surprise. Incomplete surveys were undertaken by the former Nature Conservancy during the late 1960s to identify a series of areas meriting protection as National Nature Reserves (Ratcliffe 1977b). Subsequently, the portents for land-use developments in various parts of Britain indicated that priorities for allocation of scarce survey resources should lie in other areas, with emphasis on different impacts. The launching of more comprehensive vegetational and ornithological surveys of the Caithness and Sutherland peatlands during 1979/80 has been overtaken by the rapid spread of afforestation, requiring that the conservation case be presented quickly, albeit from incomplete evidence. The uplands are the last great area of undeveloped natural and semi-natural habitat in Britain, and these northern peatlands are outstandingly valuable but especially vulnerable. This report aims to present the case for regarding the Caithness and Sutherland peatlands as both a national and an international scientific and cultural resource.

1.2 Survey methods

Since 1980 NCC has undertaken botanical surveys of the peat bogs within both administrative districts — the Peatland Survey of Northern Scotland. This has concentrated particularly on identifying and studying the patterned flows from large-scale maps and aerial photographs. The sites thus identified have subsequently been examined in detail on the ground. Such a programme of survey is inevitably time-consuming because of the nature of the terrain and the remoteness of many sites, and it is only now nearing completion (Chapter 2). Virtually all major pool systems have been examined in detail, and about 90% of the total area has been assessed at least in outline.

Complementing this survey, a series of ornithological sampling surveys have also been undertaken by NCC since 1979 — the Upland Bird Survey, continuing from 1986 as the Moorland Bird Study (Chapter 3). The practical difficulties of counting birds over such a wide expanse of peatland have meant that it has not been possible for NCC to survey all areas of the Caithness and Sutherland peatlands for their ornithological interest. The approach taken has been that of sampling representative sites throughout both districts. Thus, this report:

- briefly draws upon the results of NCC's Peatland Survey (the full results of which will be published elsewhere) in order to relate the ornithological surveys to concurrent studies of peatland vegetation and structure and to provide a habitat context for the bird fauna (Chapter 2);
- outlines the methods of data collection used by the Upland Bird Survey (UBS) in 1979-1985, and later the Moorland Bird Study (MBS), to gather information on the ornithological importance of the peatlands (Chapter 3);
- develops and assesses methods of interpolating the results of the survey to the rest of the area, using correlations between breeding densities of waders and features of physical structure and vegetation of peatland, whereby the relative importance of sites for waders can be identified both from maps and on the ground without the necessity for a full field survey (Chapter 4);
- assesses the former and current extent of peatlands and the degree to which bird populations in Caithness and Sutherland have already been reduced by commercial afforestation and then estimates potential future losses (sections 4.4 and 6.2);
- summarises the effects of afforestation on the physical and biological components of the peatland ecosystem (Chapter 6);
- assesses the international importance of the peatlands according to the requirements of the UK's treaty arrangements (Chapter 7);
- evaluates the overall biological importance of the peatlands, using information gained from both of the bird surveys, the Peatland Survey and other sources, and synthesises the total conservation case (Chapter 8).

The blanket bog 2

The following account of the blanket bog formation is based mainly on Lindsay et al. (in prep.) and included here to provide an essential context for subsequent chapters.

2.1 The Peatland Survey of Northern Scotland

During the last seven years, most of the peatlands of the north of Scotland have been systematically surveyed. In order to identify important areas of peatland, aerial photographs and 1:25,000 maps of Caithness and Sutherland were examined. A wide range of these sites was then visited to assess the present condition of those remaining free from forestry or intensive drainage. A full evaluation was made of their physical and hydrological structure and floristics. In particular, records were made of any damage to the site by drainage or previous severe burning. A high, unmodified water-table was considered to be a useful indicator of high conservation value. Sites with a high water-table were also found to display a range of natural microtopographical elements characteristic of undamaged bogs such as spongy ridges, hollows, pools and *Sphagnum* 'lawns'. Detailed descriptions of the peatland flora were made at each site and these were complemented by quantitative information on species composition within $0 \cdot 5m$ quadrats. Such data enable the plant communities of these northern bogs to be evaluated within a national context.

There are problems of definition over the limits of peatland as a substrate and as a vegetation category, since within a peatland landscape there is usually a gradual transition from deeper peat with mire communities to mineral soils with shallow surface humus, supporting drier grassland or dwarf shrub heath. An accepted though arbitrary definition according to substrate is a minimum depth of 30cm of organic deposit (Clymo 1983). This is the basis for the

Table 2.1 Plant communities of the Caithness and Sutherland blanket bogs
These were classified by McVean & Ratcliffe (1962) under the following main categories (with their table and list numbers in brackets).

Ombrogenous mire		
0.1	Trichophoreto-Eriophoretum typicum	(Table 49, lists 1-12)
0.2	Calluneto-Eriophoretum	(Table 50)
0.3	Trichophoreto-Callunetum	(Table 52, lists 1-12)
0.4	Molinieto-Callunetum	(Table 52, lists 13-21)

Types 0.1, 0.2 and 0.3 have both *Racomitrium*-rich and lichen-rich facies, while 0.1 and 0.2 also have dwarf-shrub-rich facies. Separate pool communities of *Menyanthes trifoliata* and *Eriophorum angustifolium*, with or without *Sphagnum* species, occur, mainly in pool systems with Type 0.1.

Soligenous mire		
S.1	Trichophoreto-Eriophoretum caricetosum	(Table 49, lists 13-21)
S.2	*Molinia-Myrica nodum*	(Table 53)
S.3	Sphagneto-Juncetum effusi	(Table 54, lists 1-9)
S.4	Sphagneto-Caricetum sub-alpinum	(Table 55, lists 1-9)

Types 0.1-0.4 are readily recognised and extensive: they are the principal means of defining the limits of the peatland resource. Types S.1-S.4 are common within the ombrogenous mire expanses but cover a much smaller total area.

The National Vegetation Classification being compiled at the University of Lancaster by Dr J. Rodwell for NCC has further refined this classification, but it is not yet fully available.

distribution of the main peatland areas of the map in Figure 1.2, though there are numerous other patches too small and fragmented to be shown at this scale.

Full details of the methods and results of the Peatland Survey are to be published by Lindsay *et al.* (in prep.). Their report will show the range of bog types and peatland vegetation across Caithness and Sutherland. In the interim, however, the main plant communities found in these peatlands are listed in Table 2.1.

2.2 The origin of the peatlands in relation to forest history

Many ombrotrophic bog systems first became established when the wetter conditions of the Atlantic period followed the drier Boreal period, around 5,500 BC, though some of the shallower upland blanket bogs did not begin to form until later. Many bogs show a layer of tree stumps and fallen logs within the basal peat, suggesting that forest was often overwhelmed by the rapid growth of bog-mosses (*Sphagnum* species). However, the frequent occurrence of a charcoal layer at the base of the peat has led some ecologists to infer that human activity first destroyed the forest and that the resulting increase in soil acidification was as much responsible for the growth of *Sphagnum* as the direct effect of wetter climate. Perhaps both factors acted in combination. There is good evidence from pollen analysis that some parts of northern Scotland, such as the interior of Caithness, Orkney and Shetland, never carried significant tree cover during the whole of the Post-glacial period (Peglar 1979; Moar 1969; Johansen 1975; see also Birks 1984).

It has been supposed that the spread of trees onto blanket bog surfaces represented periods of return to drier climate (notably during the Sub-boreal period from 3,000 to 500 BC), causing a desiccation of the surface and loss of *Sphagnum* and other moisture-loving vegetation. Scottish forest history is confused by the fact that tree remains, notably of Scots pine *Pinus sylvestris,* of widely varying radio-carbon age, occur at different levels in blanket bog peat in different localities. However, pine stumps in the north-west Highlands mostly date to 2,000-2,500 BC, and their occurrence beneath a deep layer of ombrogenous peat is consistent with renewal of active bog growth as a result of increased climatic wetness marking the onset of the Sub-atlantic period (Birks 1975). This climatic period, which has prevailed since about 500 BC, is not only wet, but its summer mean temperature is also believed to be cooler by 2°C than that of the Atlantic period (Godwin 1975). This has caused a descent of the tree-line by perhaps 200m and given optimum conditions for renewed growth of blanket bog. Under this present climatic regime, it is clear that trees would naturally be absent from much if not all of the deeper blanket bog surfaces in Caithness and Sutherland and that blanket bog represents a natural climatic climax type (Tansley 1939; Godwin 1975; Moore 1987).

During the Sub-atlantic period, the natural tree-line in Caithness and Sutherland has probably been at around 300m in favourable situations where woodland cover has been able to develop. It disappears altogether on the most exposed coasts, where montane heaths with species such as *Dryas octopetala* occur almost down to sea level. Before human impact began to remove it, woodland probably had a patchy cover over the region, mainly on the drier and more fertile mineral soils away from the peatlands, in the glens and on lower hillsides. Scattered remnants indicate that it was mainly birchwood *(Betula pubescens* subsp. *odorata)* of a subarctic type, with variable amounts of hazel, willow, rowan, alder and holly locally. Pinewood probably occurred locally in Sutherland, where widespread remains in the peat show that Scots pine was certainly present locally during the Post-glacial period (Crampton 1911; Birks 1975). The native trees in most places are of small size, as are the remains buried in the peat.

2.3 Peat formation and bog types

The recent history of bog development and the present state of the living surface of vegetation reflect recent and contemporary environmental conditions, especially of climate.

For nearly the last 3,000 years Britain and Ireland have been subject to a climate which is both cool and moist, derived from the North Atlantic Westerlies, which steadily gather moisture during their 4,000km crossing of the northern ocean. On passing over the first land-masses in their tracks they produce measurable precipitation as often as two days in every three over the most oceanic parts of the British Isles. Such conditions, combined with a prevalence of rugged terrain, high winds, low summer temperatures and nutrient-poverty resulting from hard, acidic rocks, severely limit the potential for agriculture and forestry in north-western Britain, tending instead to encourage the development of a vegetation dominated by plants adapted to humid and acidic conditions, and on flatter ground particularly by *Sphagnum* (Tansley 1939; Thompson 1987; Lindsay 1987; Moore 1987).

Sphagnum is a delicate plant which is easily damaged by burning, draining or even trampling. Nevertheless, almost half the 30 species of *Sphagnum* which occur in Britain are capable, under the right conditions, of producing a continuous ground layer of vegetation. The plant is also remarkable in its ability to absorb water, enabling it to maintain the ground surface in a constantly waterlogged, relatively anaerobic state. Without oxygen, the normal process of decomposition becomes inhibited and, as a result, *Sphagnum* and other plant remains fail to decompose when they die, but accumulate over the soil surface as peat instead. The major source of water and plant nutrients in such mires eventually changes from ground-water to atmospheric fall-out because the thickness of accumulated peat insulates the living vegetation from the mineral ground beneath. These are ombrotrophic (= rain-generated) mires, or true bogs, which contrast strongly

The surface of highly patterned blanket bogs shows a bewildering display of pools of different sizes and shapes, illustrating the tundra-like appearance of these wetlands. Badanloch Bog, Sutherland, August 1986

with the conditions of fen-peat formation, where sedges, 'brown' mosses, herbs and reeds are dominant under the influence of base-rich ground-water (Tansley 1939). The lack of decomposition in peat systems means that macrofossil remains and pollen grains locked in the stratigraphic profile of the deep peat deposits have been of enormous value to Quaternary ecologists in revealing the ancient vegetational history of the British Isles (Godwin 1975).

In lowland areas of England and Wales, isolated ombrotrophic peat deposits have developed as 'raised mires' in places where topography originally caused the ground water-table to remain consistently high. These are formed where earlier lakes and shallow basins have accumulated sufficient organic material to create a dome of peat (from which they derive their name) rising some 4-5m above the surrounding land. The properties of the peat maintain the dome in a state of almost constant saturation throughout the year (Ingram 1982).

In western and upland Britain, where rainfall exceeds 170 'wet days' a year (Figure 2.1) and average water balance gives a consistent surplus of precipitation over evapo-transpiration between April and September (Figure 2.2), peat develops also on plateaux and gentle inclines. Under extreme wetness, of more than 220 'wet days' a year, shallow peat occurs on slopes of up to 20 or even steeper on north-facing slopes. The deepest and wettest areas of peat, dominated by common cottongrass *Eriophorum angustifolium* and a wide range of *Sphagnum* species, tend to form on gentle slopes and level ground, whereas on steeper gradients peat is thinner and is characterised by bryophyte-rich dwarf shrub heath or acid grassland swards. These communities of shallower peat are often referred to as wet grass-heaths. The climate of these regions is so wet that peaty soils with moist heaths are extensive even on steeper and stonier ground. Vegetation characteristic of drier ground can however occur on deep peat which has suffered from burning or drainage.

The extreme wetness of the climate ensures that waterlogging occurs almost irrespective of the underlying geology. This is considered to be the ultimate development of an ombrotrophic mire system, resulting in the gradual increment of a completely organic terrain. In this way upland areas with predominantly gentle relief in northern and western parts of Britain, from Dartmoor to Shetland, have, over the last 2,000-7,000 years, become covered in a smothering mantle of peat commonly known as 'blanket bog'. The main expanses are in Wales, the Pennines, the Cheviots, the Southern Scottish Uplands, the Scottish Highlands, the Isle of Lewis and Shetland (Figure 2.3).

Although such blanket bog landscapes are widespread in western Britain and Ireland, their most extreme and extensive development is in northern Scotland. Here, to the east of the Moine Thrust (Figure 1.3), which runs from Loch Eriboll to the Sound of Sleat, the vast low-lying moorland which makes up eastern Sutherland and Caithness represents the greatest continuous area of blanket bog in Europe and has been described as unique in world terms by international peatland experts (International Mire Conservation Group 1986).

The distribution and extent of peatland are affected by the landforms of the underlying geology of the area. Most of the Caithness flows have developed on Old Red Sandstone. Further west, around Strathnaver, the bedrock of ancient Moine schists and gneisses becomes harder and less uniform in structure. West of the Moine Thrust peat development is discontinuous (Figure 1.2). Here, the ancient Lewisian Gneiss, Torridonian Sandstone and Cambrian Quartzite form a range of mountains running from Foinaven and Arkle in the north to Ben More Assynt in the south. In the extreme west and to the south, the underlying Lewisian Gneiss has been exposed and severely glacially scoured (Gordon 1981). There is a characteristic terrain of irregular low

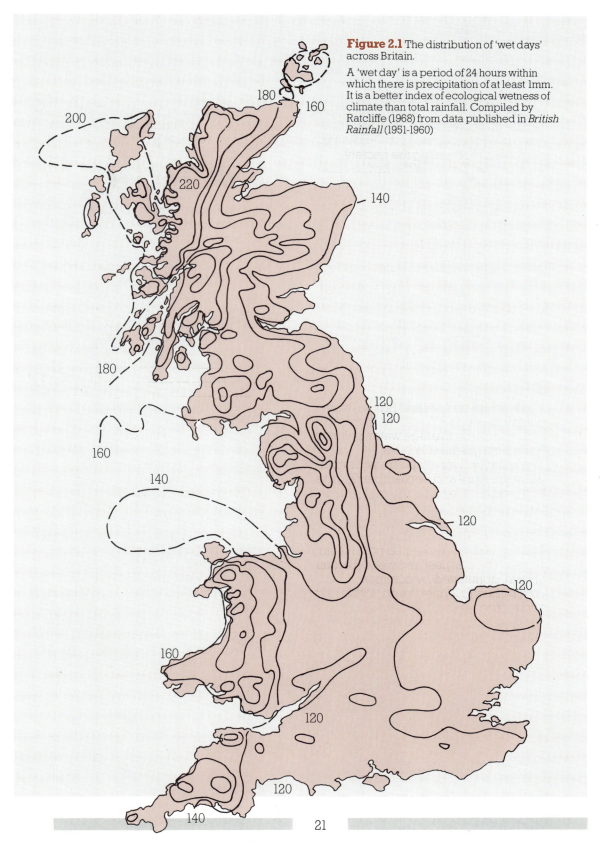

Figure 2.1 The distribution of 'wet days' across Britain.

A 'wet day' is a period of 24 hours within which there is precipitation of at least 1mm. It is a better index of ecological wetness of climate than total rainfall. Compiled by Ratcliffe (1968) from data published in *British Rainfall* (1951-1960)

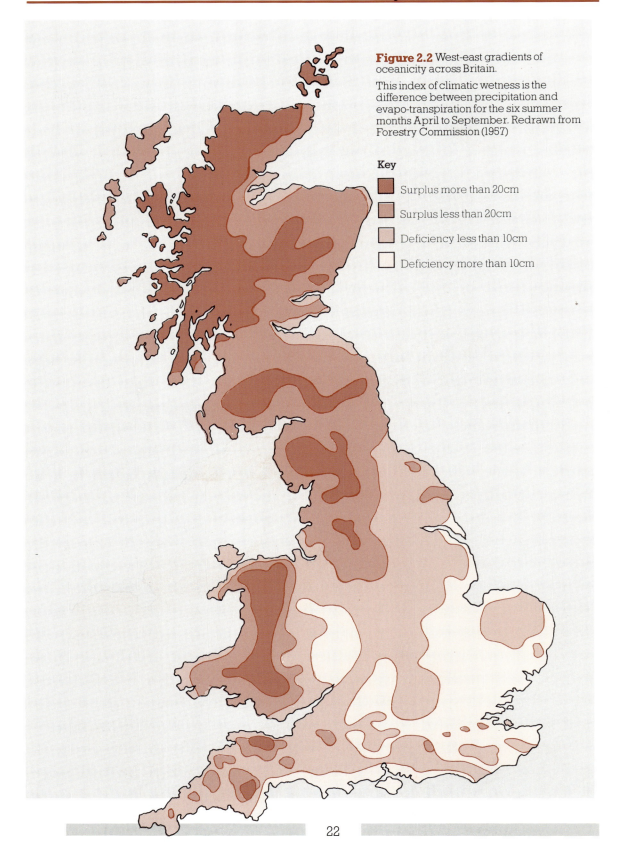

Figure 2.2 West-east gradients of oceanicity across Britain.

This index of climatic wetness is the difference between precipitation and evapo-transpiration for the six summer months April to September. Redrawn from Forestry Commission (1957)

Key
- Surplus more than 20cm
- Surplus less than 20cm
- Deficiency less than 10cm
- Deficiency more than 10cm

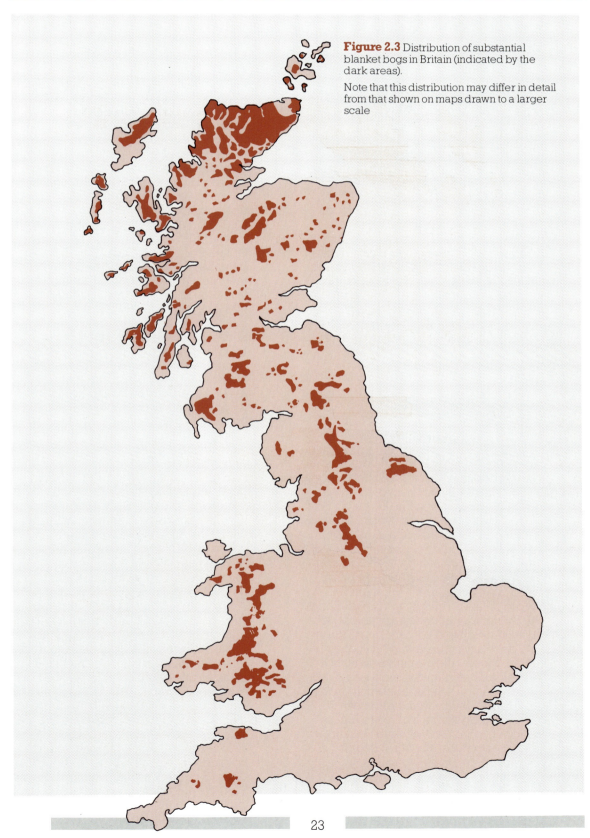

Figure 2.3 Distribution of substantial blanket bogs in Britain (indicated by the dark areas).

Note that this distribution may differ in detail from that shown on maps drawn to a larger scale

moorland, with knobbly bosses of projecting bedrock, stony moraines with wet heath, and dissected but well developed blanket and valley bog in the intervening hollows, terraces and flats.

2.4 The international distribution of blanket bog

Globally, blanket bog is an extremely rare habitat, restricted to the few areas where cool oceanic conditions prevail (Gore 1983). It is found in Europe in western Norway, the Pyrenees, Great Britain, western Ireland and Iceland, though in this last country much of the blanket bog is modified by wind-blown volcanic soil to produce a rich, fen-like vegetation (Einarsson 1968; Goodwillie 1980). Elsewhere (see Figure 1.1), it is recorded from only relatively small areas in Labrador, Alaska, Kamchatka, the Falkland Islands and Tierra del Fuego, together with small pockets in New Zealand and the Ruwenzori Mountains in Central Africa (Gore 1983).

The total global resource of blanket bog is estimated to be little more than 10,000,000 ha, of which Great Britain had between one-tenth and one-seventh (see Figure 2.3), though large areas have now been lost to forestry. The Republic of Ireland has only 771,800 ha of blanket bog, compared with Great Britain's 1,000,000 ha, and has lost an even larger proportion of its bogs through commercial peat extraction, afforestation and reclamation. Many of the most important individual areas of Irish blanket bog have already been destroyed (Reynolds 1984; Ryan & Cross 1984; van Eck *et al.* 1984; Bellamy 1986). Norway, although appearing to have widespread blanket bog (Figure 1.1), does not possess large continuous tracts. Most of the country is so rugged and steep that blanket bogs tend to be scattered in small pockets through the landscape, rather than being the dominant landform. Thus the United Kingdom contains a greater total area of blanket bog than any other country in Europe.

2.5 The significance of surface patterning

Bog surfaces are a product of organic growth largely involving a range of *Sphagnum* species, which have a variety of growth forms and are adapted to different degrees of ground wetness. A bog surface with undulations which produce a varying water-table therefore tends to display a range of small-scale patterns resulting from these various growth forms in a mosaic of *Sphagnum*. The 'hummock-hollow' pattern often referred to when describing bog systems is in fact only a small part of the full variation displayed by bog surface features. In northern Britain the 'hollows' are more likely to be water-filled pools, while 'hummocks' may vary from high moss hummocks to low, soft ridges. Lindsay, Riggall & Burd (1985) have suggested that the distribution of these surface features and their nature are determined partly by climate and partly by slope. They suggest that climate determines the maximum range of surface features for a given area, while the arrangement of these on any particular mire depends on surface gradient, as postulated by Goode (1973).

Although the flat or gently sloping ground and the high humidity and rainfall (Figure 2.1) produce blanket bog in Caithness and Sutherland which is largely of an intensely patterned type, considerable variation still exists within the mire systems of these northern districts. This variation is derived partly from the different local patterns of topography and water movement (Pearsall 1956; Boatman & Armstrong 1968) and partly from the broader regional trends resulting from altitudinal and climatic variation. Of these broader trends, the most important are the west–east variation corresponding to an oceanic–continental climatic gradient (Figure 2.2) and the south–north variation relating to the altitudinal drop from the southern hill ground to the more northerly low-lying flows.

The general range of patterns found by Lindsay, Riggall & Burd (1985) across Great Britain as a whole can be seen in Figure 2.4. Within these patterns the

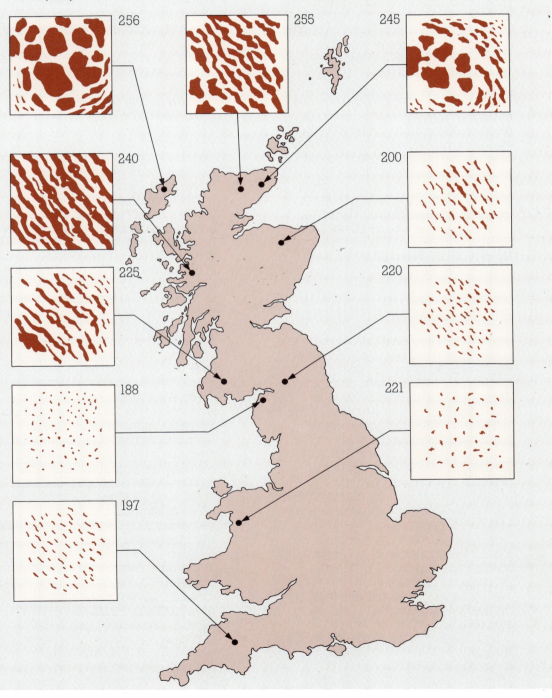

Figure 2.4 Distribution and surface structure of the pool patterns on some British bogs.

Within each box (drawn at a scale of 1:1,000), pools and hollows are shown dark, ridges and hummocks light. The adjoining numbers indicate average 'rain days' per year for each site, taken from *The Atlas of Britain and Northern Ireland* (published by Oxford University Press in 1963). From Lindsay, Riggall & Burd (1965) surface undulations are so small and the water-table fluctuations so limited that the major environmental gradients produce a series of narrow zones each determined by its vertical location in relation to the average position of the water-table. Each zone (with a vertical span of no more than 10cm) supports its own characteristic plant communities, which rely on the stability of the bog as a hydrological system for their continued survival. Lindsay, Riggall & Bignal (1983) and Lindsay, Riggall & Burd (1985) have described a range of such zones for Britain. These zones are important not just for their botanical features, but also in providing niches for the bog invertebrate fauna which, in turn, plays a major role in determining feeding sites for birds.

The range of physical patterns and their associated vegetation zones provides a significant amount of variation between bog systems throughout Britain. Sites which have largely similar species lists can still display many differences in the arrangement of their patterns and vegetation zones. Curtis & Bignal (1985) have investigated the physiognomy of peatland vegetation and shown how this varies within and between peatland sites. Vegetation structure is affected by physical patterning because hydromorphological differences result in gradients of species abundance, plant growth and thus structure.

The floristics of the blanket bogs in Caithness and Sutherland are important not only as the extreme expression of oceanic influence, compared with mire systems elsewhere in Britain, but also for the variations which occur within the region. The most pronounced trends across the two districts in both patterning and vegetation are from east to west. The increased level and frequency of precipitation towards the west results in a gradual shift from vegetation which clearly displays rain-fed (ombrotrophic) or bog characteristics to mires which begin to take on the character of Fenno-Scandian 'sloping fen'. This development occurs because the constant rainfall onto what is a naturally sloping bog type ensures a steady rate of water seepage through the surface layers almost throughout the year. Thus many valleyside flows, though generally classed as rain-fed blanket bog, support vegetation types which are more closely related to wet heath or even valley mire in southern Britain and Fenno-Scandia. Valleyside flows in the west are often characterised by an abundance of *Molinia caerulea, Myrica gale, Narthecium ossifragum, Drosera intermedia* and occasionally *Sphagnum pulchrum* — all species well known from, for example, the Dorset heaths. The abundance of *Racomitrium lanuginosum* as a component of hummock vegetation increases markedly westwards. *Drosera intermedia, Carex limosa, Schoenus nigricans, Rhynchospora alba* and Atlantic bryophytes (e.g. *Campylopus atrovirens* and *Pleurozia purpurea*) also increase in occurrence in a westerly direction.

In contrast, eastwards across the two districts the vegetation reflects continental influences, particularly in the dominance of a mixed dwarf shrub layer. Such a vegetation structure shows some affinities with the bogs of Fenno-Scandia, but certain peculiarities make the Scottish type quite distinct. Whilst in Fenno-Scandia the range of dwarf shrub species typically includes *Chamaedaphne calyculata, Ledum palustre, Vaccinium vitis-idaea, V. uliginosum, Betula nana* and *Calluna vulgaris*, in Caithness the list is shorter with *Calluna vulgaris* and *Erica tetralix* forming the major part of the shrub layer. *Chamaedaphne* and *Ledum* do not occur naturally in Britain, whilst *Vaccinium vitis-idaea* and *V. uliginosum* are restricted to steep slopes. Surprisingly, *Betula nana,* perhaps one of the most characteristic dwarf shrubs of Fenno-Scandia, is not found at its most abundant in eastern Caithness, but instead appears to favour a central position, occurring on high ridges or hummocks in undamaged mires or more generally distributed on damaged bogs. Perhaps the most peculiar feature of the dwarf shrub

Top left: Common cottongrass *Eriophorum angustifolium*
Top right: Oblong-leaved sundew *Drosera intermedia*
Centre left: Bog asphodel *Narthecium ossifragum*
Bottom right: White beak-sedge *Rhynchospora alba*
Bottom left: One of the bog-mosses *Sphagnum recurvum*

layer is *Arctostaphylos uva-ursi*. Generally described as indicative in British mires of continental influences, it is not a mire plant at all in Fenno-Scandia, but a woodland species. Similarly, *Arctostaphylos alpinus* grows as a calcifuge bog species in Britain, but on the Continent it is known as a plant of dry habitats, and in the Alps it is a strict calcicole.

Lichens of the reindeer-moss type (*Cladonia* species) also increase eastwards, though there are also western outliers of lichen-rich bog in coastal situations. These northern mires are important for their populations of nationally scarce species such as *Carex limosa, Betula nana, Arctostaphylos alpinus, Vaccinium microcarpum* and the bog-mosses *Sphagnum imbricatum, S. fuscum* and *S. pulchrum*. Blanket bog plant communities are, however, intrinsically species-poor, and their botanical interest lies mainly in the combinations of species in dynamic mosaics which represent the response to semi-aquatic conditions.

Patterned surfaces are associated with various types of mire — raised mires, aapa mires and palsa mires. Similar features are also widespread on patterned tundras in the arctic regions of the USSR, Canada and Alaska. The Sutherland and Caithness bogs represent the most southerly and oceanic occurrence of these marked patterns over a wide area and are also unusual in being developed mainly on blanket bogs. Whilst there is a superficial resemblance between British blanket bogs and certain types of low arctic wet tundras, the latter occur in areas of low precipitation, where the winter freeze produces drought conditions and the underlying permafrost causes flooding in summer. The occurrence in Britain of a naturally treeless tundra type of ecosystem far to the south of its main circumpolar distribution and at very low altitudes is also a feature of great ecological and bioclimatic interest. It illustrates how completely different sets of environmental conditions can produce a convergent response in vegetation and substrate development.

Altogether, in the total extent of blanket bog, the diversity and uniqueness of the patterned flows and the naturalness of many areas, the Caithness and Sutherland blanket bogs represent one of the most distinctive and localised of European ecosystems. To the best of our knowledge, there is no other area quite like this anywhere in the world. It is more peculiarly British than almost any other vegetation complex, except perhaps certain localised bryophyte communities of the western mountains and some types of anthropogenic grassland and heath. The area of greatest similarity, the Bog of Erris in County Mayo, has been so degraded by extensive commercial peat-working that it is no longer comparable in nature conservation importance (Bellamy 1986).

The range of surface patterns in Caithness and Sutherland and the botanical variety displayed by zones within these patterns will be reported in detail by Lindsay *et al*. (in prep.).

2.6 Threats to and losses of peatlands in the British Isles

Peatlands clearly assignable as raised bogs have always been localised in Britain. In a study to be published by NCC (Bragg *et al*. in prep.), the major original concentrations of lowland raised bog in Britain were examined for changes in land-use since the middle of the last century. Between that time and 1978, 84% of this habitat was found to have vanished through afforestation, agricultural reclamation and commercial peat-cutting; the increasing role of afforestation in this process is especially noteworthy. Much of the remaining area of raised bog has been severely damaged by burning and draining, leaving only 6% of the original 13,000 ha as still vigorously-growing *Sphagnum*-dominated bog. In total, therefore, 94% of the resource has been lost, more than half of this since 1945. If that rate of attrition is allowed to continue, the remainder will be lost in 30 years.

Blanket bog remains a far more extensive type than raised bog, though much has been lost or degraded by the same processes of change. Higher-level bogs are especially prone to damage by erosion of the peat, beginning with gullying and ending in sheet denudation. Fire and grazing have been so widespread and long continued that the proportion of the total large area of blanket bog remaining quite natural and undamaged is now quite small, and nearly all of it is in Scotland.

Blanket bog is now under intense threat in Britain, mainly from afforestation. Extensive areas have been planted in Wales, the Cheviots and the Southern Scottish Uplands. Extensive flow-lands in Wigtown district are already widely afforested. Again, 30% of the blanket peat on the Kintyre peninsula has been lost to forestry since 1945 (Nature Conservancy Council 1986), as have numerous scattered areas in both the western and the eastern Highlands. On Caithness and Sutherland peatlands about 67,000 ha are planted or programmed for planting, and, even though not all of this is planted yet, afforestation has been so scattered that only eight out of 41 hydrological systems remain free from some planting. Planting has taken place up to the edge of many patterend bog systems.

Ireland, another major stronghold of European blanket bog, is also seeing a marked decline in this habitat. Indeed, bog systems generally are under great threat in Ireland. Ryan & Cross (1984) quantified the rates of exploitation for all Irish peatland types. They found that blanket bogs were less damaged or modified than other peat bog types, yet, even so, some 207,900 ha had been damaged. This amounts to 27% of the total blanket bog in Ireland, while in all nearly 50% of Irish peatlands have been lost as natural ecosystems. Such loss has been piecemeal with no planned policy for conservation of key sites (van Eck *et al.* 1984; Bellamy 1986).

There have been and there still are plans to extract peat from Caithness and Sutherland both for fuel and for horticultural uses. The area contains vast reserves of deep peat, and the Scottish Peat Committee set up in 1949 was especially interested in them (Scottish Peat Committee 1968), but concluded in 1962 that these deposits were not worth working commercially under the economic conditions then prevailing. Recently, there has been renewed interest in the possibilities of large-scale peat-working, but there is no reason to suppose that economic constraints have changed significantly, notably the distance from markets. So far the extraction is mostly local and small-scale, around the edges of the main peatland masses. However, the possibility of major EEC funding for commercial peat extraction (*The Scotsman,* 5 December 1986) may affect the economics of working peat in Caithness and Sutherland.

NCC's Upland Bird Survey in Caithness and Sutherland

3.1 Introduction

The mountains and moorlands of Britain have an ecologically diverse and distinctive bird fauna containing outlying and insular populations of species which belong either to high altitudes or to northern latitudes in continental Europe. Of the various types of upland habitat, the wet moorlands with large expanses of blanket bog are a particular feature of the highly oceanic British climate, and their associated bird fauna is also of outstanding interest. The bird community of these northern peatlands is especially rich in species of waders and represents an unusual avifauna, showing affinities with those of both boreal mires and arctic tundra.

Blanket bog is a naturally treeless ecosystem lying latitudinally within the boreal and cool temperate forest zones. Despite certain resemblances, it differs from arctic tundra in that its origins depend not on permafrost but on a cool and extremely humid climate. In Europe, blanket bog reaches its greatest development in the far north of Scotland, where its extent and variety of form represent one of the most remarkable vegetational features of Great Britain. Our concern has thus been to record and evaluate the ornithological character of this important habitat in parallel with survey and assessment of its vegetation.

Earlier writers on ornithology made brief reference to the breeding bird fauna of the boggy flat moorlands ('flows') of east Sutherland and Caithness. Harvie-Brown & Buckley (1887, 1895) describe the great flow land of this district, with its numerous dubh lochans, as an important nesting haunt of red-throated diver, greylag goose, greenshank, golden plover, dunlin, red grouse, wigeon and five species of gull. During the earlier part of the present century, the area appears to have been little visited by ornithologists, but a few egg-collectors came in search of the rarer species. It became known as perhaps the main breeding area of the common scoter in this country. In 1900 and 1901, E. S. Steward, a notable oologist, recorded nesting greylags, wigeon, common scoters, red-breasted mergansers, greenshanks, red- and black-throated divers, common gulls and arctic skuas from the flows around Forsinard. In a vivid description of the Caithness flows, Yeates (1948) drew attention to the occurrence here of small mainland colonies of arctic skuas, whilst Rankin (1947) discussed the breeding black-throated divers; both authors mentioned some of the other species emphasised by Harvie-Brown & Buckley. Assessment of conservation value came much later. Ratcliffe (1977a) regarded the Caithness and Sutherland flows as the most important blanket bogs in Britain for variety of bird species, and Fuller (1982) stressed the unusualness of the bird assemblage of these northern moorlands in his study of British bird habitats.

By 1970, the conservation importance of the Caithness and Sutherland peatlands, both for vegetational features and for birds, was realised and an initial selection of the most important areas suggested. While it was recognised that this choice was based on extremely fragmentary and qualitative survey information, it was evident that some conservation measures had to be achieved whilst further surveys were undertaken. By the late 1970s, there were clear portents that transformation of these great blanket bog systems on a massive scale, through afforestation, was becoming ever more likely. Moreover, while quality of the best areas for breeding birds was assessed largely according to species diversity and population density of notable species, it had become clear that the total population sizes for some species had both national and international importance. This gave a new dimension to the survey and evaluation of the ornithological interest of the northern Scottish peatlands, as the basis for their conservation.

In 1979, NCC launched a programme of breeding bird surveys of moorlands likely to be affected by afforestation, and this has particularly concentrated on Caithness and Sutherland. Its aims were fivefold:

- to identify, from sites surveyed, breeding bird assemblages of high nature conservation interest in terms of species diversity and population density;
- to collect data to assess and identify habitat features important to the breeding birds and from these associations to predict the location of other areas of high ornithological interest;
- to estimate the size of the populations of breeding birds (especially waders) dependent on the peat flows of Caithness and Sutherland;
- to understand the effects that rapid changes — especially those resulting from afforestation — would have on the birdlife of these wetlands;
- to make recommendations for the conservation of wetland bird assemblages within Caithness and Sutherland.

To these ends, sample areas were surveyed in Caithness in 1979, 1980 and 1984, and in Sutherland from 1980 to 1986. The results of these surveys have been or will be published elsewhere (Symonds 1981; Langslow & Reed 1985; Reed & Langslow 1985, in press and in prep.; Reed, Langslow & Symonds 1983a, 1983b; Barrett *et al.* in prep.).

Additionally to NCC's survey work, the Royal Society for the Protection of Birds (RSPB) carried out other studies within the same region between 1980 and 1986. Slightly different methods were used to allow larger areas to be covered at the expense of fine-scale detail. This work has extended the area for which data are available and permitted an independent assessment of the breeding bird assemblages of some areas. RSPB's results have been presented as a series of internal reports. In 1985 RSPB surveyed eight sites in Caithness and Sutherland using NCC's methodology (Birkin, Hayhow & Campbell 1985). The results of this survey have been used to extend the scope of the NCC data-base.

NCC and RSPB quantitatively surveyed waders on a total of 77 sites (sample areas) in Caithness and Sutherland. The range of size of the sites was 200-1,025 ha. Despite this large range, over 80% of sites (62 of 77) lay in the range 550-950 ha, with sites clustered around the mean value of 674 ha ($\pm$ 174 sd). Most sites were surveyed in only one year but some were visited in two or more seasons. A few were surveyed for up to five consecutive years in order to investigate whether there were medium-term population changes. Whilst in the early years of the survey the emphasis was placed on recording waders and wildfowl, more recently all species of birds encountered on survey sites have been recorded.

The selection of sites was made so as to include those thought likely to be important as well as others giving a range of quality typical of the region as a whole. The sites were chosen to include many examples of all the major peatland habitats (see Chapter 2) within the total range of types occurring in Caithness and Sutherland. Thus one of the aims of the project has been to assess typical breeding densities over the whole of the peatlands, and not just on the prime sites with exceptional densities. By relating varying density to readily surveyed ecological features, it then becomes possible to estimate densities in other parts of the region which have not been surveyed for birds. In this way, the total size of the breeding populations in the region can be assessed with some degree of confidence, and individual areas placed in context.

3.2 Programme of ornithological surveys

Methods

The method used was a modification of the territory-mapping census. A detailed description of methods used and tests employed to ensure consistent and accurate recording and interpretation of results is presented in the Appendix.

Distribution of sites surveyed

The location of sites surveyed by NCC between 1979 and 1986 and by RSPB in 1985 is shown in Figure 3.1. In Caithness, most sites were located in the centre of the district, although two sites were

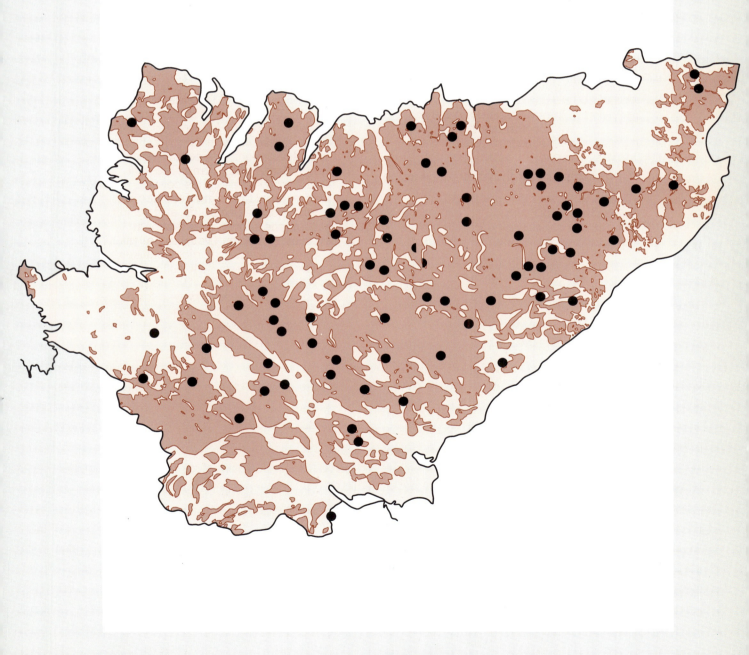

Figure 3.1 Distribution of sites surveyed for breeding waders (1979-1986) superimposed on the blanket bog shown in Figure 1.2.

Eight sites surveyed by RSPB in 1985, using NCC methods, are included

Key

▬ Blanket bog
● Ornithological survey sites

surveyed in the north. In Sutherland, most sites were located in the east and centre, but within this area sites were geographically widespread and covered the full range of peatland habitat types. Indeed, across both districts the sites surveyed ranged from coast to coast. Montane habitats and steeply sloping moorland at high altitudes were generally not included in the sample of habitats surveyed. This will mean that the sample of sites does not contain a representative sample of those species, such as ring ouzel and wheatear, which favour such ground. Separate RSPB/NCC surveys had previously been made of two of the most important birds of these steep and rocky upland habitats, the peregrine and the golden eagle. There were a few sites in which montane or steep land was included. These confirmed the general absence of waders and wildfowl from these habitats.

3.3 Composition of the peatland breeding bird fauna

The bird fauna in spring and summer includes several taxonomic/ecological groups — waders, other waterfowl, raptors and scavengers, passerines and a miscellany of other types. The waders are outstanding in overall numbers and diversity, and the occurrence of at least 15 breeding species reflects the variety of peatland and open water habitats. Golden plover, dunlin, greenshank and curlew occur with high constancy in the sample areas and so have large total populations within the region. Common sandpiper and snipe are also widespread and numerous, but several species are somewhat local — lapwing, oystercatcher, redshank and ringed plover.

The overall densities of waders at different sites varied considerably (Figure 3.2). In Caithness, densities for all breeding waders ranged from 0·9 to 14·0 pairs/km^2, and in Sutherland from 0·2 to 14·3 pairs/km^2. The overall mean density of breeding waders was 5·4 pairs/km^2, but this excludes steep, montane and other areas unsuitable for waders, as explained in Chapter 4. The number of wader species breeding on particular sites varied from one to ten.

Figure 3.2 Densities of all breeding waders on the survey sites in Caithness and Sutherland

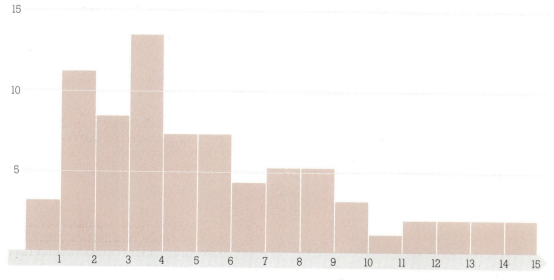

Frequency of sites

Densities of breeding waders (pairs/km^2)

In Scandinavia, several of the breeding birds of open peatland belong to the boreal—subarctic forest zone rather than to the treeless arctic tundras — greenshank, wigeon and three British rarities, wood sandpiper, ruff and Temminck's stint. They are, however, associated with forest lakes and open marshes, in which there is usually a considerable area of transitional habitat consisting of very open woodland with increasingly stunted and checked trees as the wetness of the ground increases (Sammalisto 1957; Moen 1985). Such habitats were once well represented in the Spey Valley pinewoods and supported a small and isolated population of greenshanks. The dense closed plantations of the new British forests do not provide these transitional open forest habitats on anything but a temporary and fragmentary scale. Nethersole-Thompson & Watson (1974) attributed the disappearance of the greenshank population of the Speyside pinewoods to commercial afforestation of its main breeding and feeding habitats, open heaths and marshes within the forest area (see section 6.1).

Whilst golden plovers in Scandinavia are mainly birds of open tundra, they also have a forest heath and bog niche and they formerly bred sparingly in such habitats on both Speyside and Deeside. They now appear to have disappeared from such habitats in Scotland. Even where former breeding places have been left as unplanted enclaves of up to 1 km in diameter within Kielder Forest, most of the golden plovers have gone: the ungrazed and unburned vegetation is now too tall, quite apart from any possible influence of predators.

Some bird species show interesting ecological adaptations to the British environment, differing from those characteristic of populations throughout their larger Eurasian range. The golden plover population in Britain is the southernmost representative of a circumpolar species-complex, and it has developed an interesting cline in plumage, a partial migration pattern and a local dependence on anthropogenic habitats for both feeding and breeding. The greenshank and wood sandpiper have adapted to completely treeless open moorland habitats in Scotland, though they do sometimes also occur in open bogs in Scandinavia (Hakala 1971). The recent discovery of ruff breeding in the Caithness and Sutherland peat bogs is highly unusual, since this species is usually found in sedge marshes, often within forest, in more northern areas of Fenno-Scandia, or in unimproved agricultural grasslands in the European low countries. This is also the first known breeding of ruff in Britain evidently from the arctic—subarctic population, for the recent colonisation of a few English localities appears to be from the temperate European population. These are examples of ecological divergence which are of considerable interest to students of evolution and speciation, since this is one of the ways in which new forms of organisms evolve.

The occurrence of a notable group of northern wading birds as 'fringe' species in this part of Scotland — wood sandpiper, ruff, Temminck's stint and red-necked phalarope (D. & M. Nethersole-Thompson 1986) — is also of much interest as a 'biological barometer' of climatic change. They are all still rare, but it is supposed that their tendency to appear in Britain is the response to an increase in the incidence of 'northern' climatic conditions in the form of colder springs during the last two or three decades.

Waterfowl are associated especially with open waters varying in size from tiny dubh lochans to large lochs. They include black-throated and red-throated divers, greylag goose, common scoter, wigeon, teal, mallard and red-breasted merganser. Common and black-headed gulls nest in small colonies.

At some sites predators and scavengers were regularly seen, with golden eagle, hen harrier, merlin, peregrine, short-eared owl and raven present. Some pairs of merlins, hen harriers and short-eared owls breed on the peatlands, but the other species nest on rocks and hunt over the bogs. The arctic skua is a different kind of predator breeding in small groups on the flows. The methods used here were not appropriate for the accurate calculation of raptor breeding densities, but attention is drawn to the general importance of this area for raptors (Table 8.1 and Figures 3.4f-i).

The remaining species are mostly those with a fairly widespread distribution on moorland in Britain. They include red grouse, meadow pipit, skylark and, as streamside nesters, dipper and grey wagtail.

Densities for individual species of waders were calculated in two ways. For those which occurred throughout the peatlands and were found on a majority of survey sites (golden plover, dunlin and greenshank), the mean densities were calculated for all sites surveyed. Thus the mean includes zero values for the few sites where birds were not present. For more local species of waders occurring on few sites, means were calculated in two ways — first as described above and secondly only for those sites holding breeding birds. This latter value represents local abundance where the species is present. For widespread waders such as golden plover, dunlin and greenshank the two methods give fairly similar densities.

3.4 Waders

Golden plover

Golden plovers were the most abundant breeding wader found on the peatlands of Caithness and Sutherland and were present on all but five of the 77 sites surveyed.

The range of breeding densities of golden plovers is shown in Figure 3.3a. Mean density on the sites surveyed was 1.76 pairs/km^2 (± 1.22 sd). The breeding distribution of golden plovers in Caithness and Sutherland is shown in Figure 3.4a.

When overall densities for survey sites in any year are calculated, there are slight annual differences. These differences may reflect year-to-year differences in factors affecting the breeding population such as spring weather (cf Thompson, Thompson & Nethersole-Thompson 1986) or may reflect slight differences in the average quality of sites selected for survey in each year. Low densities of 0.9 pairs/km^2 were reported by D. & M. Nethersole-Thompson (1986) over their study area of 3,250 ha in north-west Sutherland, although in south-east Sutherland in the 1960s they found that densities were higher, with approximately 2-3 pairs/km^2.

Golden plover

Breeding grounds vary widely, from wet pool and hummock flows, through drier cottongrass and deergrass bog (Trichophoreto-Eriophoretum typicum of Table 2.1) to dry heather moor, grassland and stony moraines. Small numbers of golden plovers also nest on the high plateau blanket bogs, grasslands and dwarf shrub and moss heaths within the montane zone. The common feature is very short vegetation. Golden plovers often feed in richer wet places or on grassland on the moor, but in some areas they regularly resort to the improved pastures of enclosed fields beyond the moorland edge (Ratcliffe 1976).

Dunlin

Dunlins were the second most abundant breeding wader, being found at 71 of the 77 sites surveyed in Caithness and Sutherland. Their breeding distribution over the whole of the two districts is shown in Figure 3.4b.

The range of densities of dunlin on the peatlands is highly skewed (Figure 3.3b). The density distribution has a long right tail, since a few sites hold very high densities compared with those over much of the rest of these blanket bog areas.

Dunlin

The mean density was $2 \cdot 39$ pairs/km^2 ($\pm 2 \cdot 49$ sd). In view of the skewed breeding density distribution, which is a function of this species' semi-colonial breeding habits, the median density ($1 \cdot 76$ pairs/km^2) may be a more biologically meaningful measure of abundance. As with golden plover, there were year-to-year variations in breeding density. These year-to-year differences are probably largely due to inclusion or exclusion of a small number of high-density breeding sites in the sample surveyed. Annual variations in weather may also be important in affecting breeding densities.

D. & M. Nethersole-Thompson (1986) reported densities ranging between $0 \cdot 62$ and $2 \cdot 7$ pairs/km^2 on peatlands in Sutherland, with exceptional breeding densities of up to 25 pairs/km^2 over small areas (five pairs in 20 ha). Their range of $2 \cdot 04 - 2 \cdot 70$ pairs/km^2 over nearly twenty years in one study area of 740 ha equates well with average densities found in this study.

Holmes (1966, 1970) investigated dunlin feeding ecology and breeding density on arctic and subarctic tundra. He found that breeding density related closely to the abundance of invertebrate food, which in turn was determined by habitat features including the number and spacing of pools. Holmes (1970) concluded "that the density of breeding dunlin is related to the abundance and availability of their food supply and that the main function of territorial behaviour is to disperse the populations in relation to food". This is most probably also the case on Scottish peatlands, since breeding density is strongly determined by the abundance of dubh lochans for feeding (section 4.1).

Other previous studies of dunlin have shown them to be strongly site-faithful (Soikkeli 1970; D. B. A. Thompson pers. comm.).

Greenshank

Greenshanks were the third most abundant breeding wader on the peatlands of Caithness and Sutherland and were found on 58 of the 77 sites surveyed. Their breeding distribution in the two districts is shown in Figure 3.4c.

The range of breeding densities of greenshanks on the survey sites was also skewed (Figure 3.3c). The mean density was $0 \cdot 31$ pairs/km^2 ($\pm 0 \cdot 29$ sd), whilst the median density was $0 \cdot 21$ pairs/km^2. The range of densities found is similar to that described by D. & M. Nethersole-Thompson (1986) in studies in the north of Scotland. In one study area in north-west Sutherland they found greenshank densities varying between $0 \cdot 2$ and $0 \cdot 9$ pairs/km^2 over an 18-year period.

Greenshank

There seemed to be relatively little year-to-year variation in breeding densities. Such variation as there was can be explained by the known effects of spring weather on breeding numbers and success (Thompson, Thompson & Nethersole-Thompson 1986).

Greenshanks show a more complex use of the peatland habitats than some of the other breeding waders. The species requires semi-aquatic food-gathering areas either along fairly productive rivers or at the margins of lochs and pools. However, nesting may occur at a distance of up to 3 km away, on either dry or boggy moorland which may even contain rock outcrops. During the nesting period, adult birds fly between feeding and nesting areas, and, after hatching, the young are led down to the wet feeding areas, which are rich in invertebrates.

Thus greenshanks favour gently contoured peatlands in areas where lochs and pools are plentiful, and, where available, habitats range from patterned blanket bogs to shallow valley bogs within drier morainic *Calluna* heaths. In the far west of Sutherland, nesting can occur in areas of shallow peat strewn with glacial erratics (D. & M. Nethersole-Thompson 1979, 1986).

Curlew

Curlews were the fourth most abundant species of breeding wader on peatlands in Caithness and Sutherland and were found at 46 of the 77 sites surveyed (Figure 3.3d). Their mean density was found to be 0.51 pairs/km^2 (± 0.41 sd) on those sites where they were present and 0.31 pairs/km^2 (± 0.41 sd) over all 77 sites.

Curlew

Breeding areas include a range of moorland habitats from deep bogs to dry heath and grassland, but curlews are often more numerous on the rough and often rushy enclosed pastures of the marginal and crofting lands.

Common sandpiper

Although a species associated with loch-edge and riverine habitats, common sandpipers were recorded breeding on 41 of the 77 study sites (Figure 3.3e).

Breeding densities for Caithness and Sutherland were 0.39 pairs/km^2 (± 0.35 sd) on those sites where breeding birds were present, but for all survey sites combined the mean density recorded was 0.21 pairs/km^2 (± 0.32 sd). However, this is probably an underestimate. Common sandpipers breed mainly along watercourses, and the survey methods are less efficient in this discrete habitat than over the expanse of blanket bog. It would perhaps be more valid to express abundance of this species with respect to the total length of suitable water's-edge habitat (cf D. & M. Nethersole-Thompson 1986).

Common sandpipers are highly faithful to territories where they have nested successfully in previous years: in a study in the Peak District over 85% of a colour-ringed sample returned to the same sites in successive years (Holland, Robson & Yalden 1982).

Other wader species

After the species listed above, breeding wader species in order of frequency of breeding records on the survey sites were snipe, lapwing, redshank, ringed plover and oystercatcher.

Snipe are partial migrants, widespread in the peatland areas of Caithness and Sutherland. The breeding distribution and abundance of snipe is poorly known owing to the very secretive habits of the species. Snipe were found at 45 sites, but breeding was proved at only 27 (Figure 3.3f). However, the territory-mapping methods used undoubtedly failed to locate many birds. The survey detected an average of 0·31 pairs/km² (± 0·29 sd) on those sites where birds were proved to breed, but for all survey sites combined the mean breeding density was 0·11 pairs/km² (± 0·22 sd). As with other species, the distribution of snipe was found to be localised within sites and strongly determined by the distribution of suitable feeding habitat — wet, rank flushes with abundant cover of *Juncus* species. D. & M. Nethersole-Thompson (1986) found densities ranging between 0·18 and 0·31 pairs/km² in a 3,250 ha study area in north-west Sutherland. They considered that this underestimated the abundance because birds favoured acid grassland areas within the peatlands. The major problems of surveying breeding snipe in other wetland habitats have been investigated by Green (1985).

Lapwings are partial migrants and are widespread, although commoner in Caithness than in Sutherland. In winter, lapwings leave inland and upland areas to winter on the coasts and in the lowlands. Lapwings bred on 23 of the 77 sites surveyed (Figure 3.3g). They are mainly a species of grassland and arable and not strictly a peatland wader, and they showed a strong affinity for those parts of sites with marginal agricultural improvement, either through drainage or as a result of reversion of hill pastures. The mean density on sites where they occurred was 0·66 pairs/km² (± 0·63 sd), whilst over all 77 sites the mean density was 0·20 pairs/km² (± 0·45 sd). The highest density recorded was of 1·92 pairs/km² on a peatland site in Caithness.

Redshanks are partial migrants to the peatland areas of Caithness and Sutherland, with a somewhat localised distribution within the two districts owing to their habitat requirements. They were found to be commonest in Caithness, with fewer birds using peatlands in Sutherland. Redshanks bred on only 14 of the survey sites (Figure 3.3h) and were present on five other sites. Densities were generally low. The mean density on sites where they occurred was 0·30 pairs/km² (± 0·17 sd), whilst that over all survey sites was only 0·06 pairs/km² (± 0·14 sd).

Ringed plovers are also partial migrants to the peatlands of Caithness and Sutherland. Their breeding distribution is widespread and includes inland areas. Ringed plovers were found breeding on 12 sites (Figure 3.3i), with a mean breeding density on those sites that were occupied of 0·29 pairs/km² (± 0·23 sd). The mean density over all surveyed sites was 0·05 pairs/km² (± 0·14 sd).

Oystercatchers are summer visitors to inland breeding areas within Caithness and Sutherland. They have a widespread breeding distribution, but there is little information on breeding numbers within the two districts. Oystercatchers were found breeding on only seven of the survey sites (Figure 3.3j). The mean density on these sites was 0·26 pairs/km² (± 0·16 sd), whilst over all sites it was 0·02 pairs/km² (± 0·09 sd).

The woodcock is a secretive wader with a localised breeding distribution. Though it is primarily associated with woodlands, single pairs were recorded as nesting on two sites and birds have been recorded as present on two further sites during the last decade.

Wood sandpiper

The wood sandpiper occurs in Caithness and Sutherland as a rare summer visitor and breeding species (D. & M. Nethersole-Thompson 1986). It is nationally rare, with never more than 10 pairs recorded as breeding at one time in Britain. Wood sandpipers are thought to occur regularly in small numbers on the peatlands, the vast extent of which results in few confirmed records; however D. & M. Nethersole-Thompson (1986) reported two breeding pairs on the same

Sutherland moss in 1968. During the surveys the species was recorded from two sites in Sutherland.

Red-necked phalaropes breed irregularly within the area, but the species has declined in numbers as a breeding bird in Britain and Ireland and its main stronghold is now in Shetland. A substantial proportion of one 1981 breeding site in the Caithness and Sutherland peatlands was later ploughed for afforestation. No birds were seen during the surveys.

Temminck's stint is a very rare breeding visitor to the peatlands of Caithness and Sutherland. It was not recorded from the survey sites, although it is known to have bred within the area.

The ruff is a northern breeding wader, with very small numbers breeding in the Low Countries and in England. Of particular interest is a recent record of confirmed breeding by ruff on the Caithness and Sutherland peatlands. None were seen on survey sites in the period 1979-1986.

Pectoral sandpipers have been seen with increasing frequency in Scotland in recent years (Thom 1986). A male of this Siberian—North American species displayed over a flow in Caithness during the spring of 1974 (Byrne & Mackenzie-Grieve 1974).

3.5 Other waterfowl
Red-throated diver
Red-throated divers have a boreal—high arctic world distribution and breed throughout the peatlands of Caithness and Sutherland, moving to the coast and then southwards in winter. Although there has never been a comprehensive summer survey of numbers, Thom (1986) suggests that "substantial numbers", probably up to 200 pairs, breed in Caithness and Sutherland. This amounts to the major part of the mainland breeding population. Other mainland areas, such as Argyll, hold smaller numbers (Broad, Seddon & Stroud 1986), whilst the greater part of the British population breeds on the islands of Shetland (700 pairs), Orkney (67-80 pairs) and the Outer Hebrides (39-46 pairs) (Gomersall, Morton & Wynde 1984).

Red-throated divers were present on half the sites surveyed, with two pairs breeding on one site, one pair breeding on each of a further 19 sites and birds present but with breeding unconfirmed on 15 further sites. Breeding occurs on very small lochans or pools, and birds fly to feed on either large lochs or the sea. With these nesting habits, it is possible for red-throated divers to breed in areas of highly patterned mire where there are no extensive areas of open water.

Red-throated diver

Thom (1986) records the considerable extension of the breeding range in the early 20th century, and numbers are still increasing in some areas.

Black-throated diver
The black-throated diver is a rare species in Britain whose stronghold is in northern Scotland. Elsewhere, it has a strongly northern distribution extending to the Arctic. Breeding throughout Caithness and Sutherland, black-throats select large lochs, for both nest site and feeding area (in contrast to red-throated divers) during the breeding season. Lochs with remote islands are particularly favoured, so that many lochs which are otherwise suitable but without islands are not used. Campbell & Talbot (1987) document the very low breeding success of the British population, and low productivity has been recorded for black-throated divers in Sutherland in

previous years (Bundy 1979). An important component of this failure is fluctuation in water-level of the breeding loch (Rankin 1947; Dennis 1976). Modification of the catchment hydrology owing to peatland drainage can be expected to reduce the time lag between precipitation and run-off and thus increase water-level fluctuations.

Black-throated diver

Similarly, some traditional non-peatland nesting lochs in more lowland areas have probably been rendered less suitable owing to lowland hydro-electric or water extraction schemes. Thus remote peatland lochs could have become more important for this species owing to their lack of water engineering schemes or other hydrological modification.

The total British population is estimated at 150 pairs, of which 38 were found in Caithness and Sutherland (Campbell & Talbot 1987). Black-throated divers were present on 23 survey sites, with two pairs on one site, one pair on each of eight other sites and birds present but with no evidence of breeding on 14 further sites.

With low productivity, small numbers and a restricted distribution, the British population of black-throated divers is at considerable risk. RSPB has recently initiated a major programme of research into the breeding and ecology of this species in the north of Scotland.

Little grebe

The little grebe is a rare breeding bird in the peatlands of Caithness and Sutherland, found nesting on only one survey site. It requires shallow lochs with abundant submerged vegetation for feeding and sufficient dense emergent vegetation for nesting cover. Hence, most areas of ombrotrophic peatland and dubh lochans are unsuitable, and little grebes appear to prefer areas where there is some natural nutrient enrichment giving a productive aquatic flora.

Slavonian grebe

The Slavonian grebe is a very rare breeding summer visitor to Caithness and Sutherland. Thom (1986) records its breeding in Sutherland first in 1929 (four pairs) and then irregularly until the 1960s. In Caithness, first breeding also occurred in 1929, with up to 10 pairs recorded until at least 1975, although none have been recorded since. The Slavonian grebe, like the little grebe, prefers enriched conditions and is thus not found on the more oligotrophic waters of the extensive peatlands. This species was not recorded on the sites surveyed.

Grey heron

The grey heron is a scarce resident breeding species in Caithness and Sutherland, although some individuals, particularly young birds, leave the region in winter. Here they are at the north-western limit of their Palaearctic breeding distribution. Birds were present on 14 survey sites, with breeding by one pair recorded on each of two sites. The 1954 census found two heronries in Caithness, with 12 pairs, and seven heronries in Sutherland, with 33-34 pairs (Thom 1986). More recent information suggests no major change in status within the area (M. Marquiss pers. comm.).

Whooper swan

Whilst primarily winter visitors from Iceland to large lowland lochs in Caithness, birds occasionally breed in Britain and not infrequently summer on Caithness and Sutherland peatlands. Whooper swans were present on two sites surveyed by NCC teams.

Pink-footed goose

Pink-footed geese are winter visitors from Iceland, with a major wintering locality close to south-east Sutherland

(Owen, Atkinson-Willes & Salmon 1986). Birds were present on two survey sites during the summer but there was no evidence to suggest that these birds were anything but delayed migrants.

Greenland white-fronted goose
Greenland whitefronts have traditionally fed and roosted on peatlands throughout their world range, and their present winter distribution in north and west Scotland, Wales and Ireland closely reflects the distribution of oceanic blanket bogs and raised bogs. Their status in Britain has been the subject of recent research (e.g. Stroud 1985; Greenland White-fronted Goose Study 1986).

Laybourne & Fox (in press) summarise the past and present status of wintering Greenland white-fronted geese in Caithness. These geese have been present on the peatlands since at least the 1880s. Harvie-Brown & Buckley (1887) recorded that the keeper at Strathmore bred pinioned birds from wounded geese shot on the peatlands around the Lodge. More recently, significant numbers have wintered in the agricultural lowlands, but geese still use the peatland areas for roosting and feeding, especially in autumn.

Laybourne & Fox (in press) point out that the bog-feeding Greenland whitefronts represent the only European goose still wintering on natural (rather than agricultural or semi-natural) habitat. Considering that "conservation is not merely concerned with the establishment of reserves, but is the maintenance of natural diversity", they state that the protection of these peatland areas for Greenland whitefronts is of critical importance.

There is recent evidence that peatlands in Sutherland are also used by this race of geese, at least at times in the winter (Greenland White-fronted Goose Study 1986), but further survey is required to establish the full extent of their occurrences in this district.

Greylag goose
In contrast to most breeding greylag geese in Britain, those of north-west Scotland are considered to be native in origin (Thom 1986). There are few good data on numbers of breeding greylags, but Thom (1986) refers to 11 pairs found in Caithness in 1977 and considers the whole native British stock (including those in the Western Isles) to be between 2,500 and 3,000 birds, or 500 to 700 breeding pairs.

Greylag goose

NCC surveys found three pairs breeding at one site, two pairs breeding at another site and one pair breeding on each of a further five sites. Birds were present at 27 further sites, although many of these records relate to flocks of moulting birds in midsummer. In 1986, a large moulting flock was discovered on Loch Loyal. By early July, numbers had increased to about 1,200 birds, which must represent a major proportion of the native stock breeding in Sutherland and possibly Caithness. This moulting aggregation is of significant conservation importance, and clearly further study is required to establish the provenance of these birds.

Wigeon
Wigeon breed in both Caithness and Sutherland (Figure 3.4d) and also winter within the region, though wintering birds are largely confined to the large lowland lochs of Caithness and south-east Sutherland. Breeding pairs are thinly distributed across peatland areas. Sharrock (1976) suggested that the Scottish breeding population amounted to about 400 pairs.

Of the 72 sites surveyed for wildfowl, 29 (40%) held wigeon. One site held six breeding pairs, a further 15 sites held from one to three pairs, and birds were recorded as present on another 13 sites. Their secretive nature during the breeding season means that they were undoubtedly under-recorded.

Teal
Teal both breed and occur as winter visitors in Caithness and Sutherland. The very secretive nature of breeding teal has meant that information on numbers and distribution within the region is scarce. Whilst widespread, they appear to be nowhere common and decrease in abundance in west Sutherland.

Birds were recorded from three-quarters of the survey sites (53 of 72), making teal the commonest recorded wildfowl species on these peatlands. Between four and seven pairs were recorded as breeding on six sites and one to three pairs on 27 further sites, and birds were recorded, although with no evidence of breeding, on 20 further sites.

Fox (1986a) gave details of the breeding ecology of teal on a Welsh peatland, showing that position of nests is closely determined by the location of standing water and demonstrating the importance of open water areas within peatland for feeding and brood-rearing. Areas of peatland with extensive pools, such as the highly patterned peatlands under consideration, can be supposed to be highly attractive nesting habitat for teal. The blocking of large ditches cut within the peatland, with a resultant increase in open water area, was shown to increase the numbers of teal nesting successfully (Fox 1986a). It can thus be inferred that drainage of peatlands, especially of pool systems, will be extremely damaging to the quality of teal nesting habitat.

Mallard
The mallard is both a resident breeding bird and a winter visitor to Caithness and Sutherland. Ecologically, mallard are adaptable and can utilise a wide range of freshwater habitats. Thus, unlike teal and wigeon, they are not confined to predominantly peatland areas within Caithness and Sutherland. There are no reliable estimates for breeding numbers of mallard within the peatlands.

After teal, mallard were the second most commonly recorded wildfowl species, being found on two-thirds of the sites (49 of 72). Between one and three pairs bred on 30 sites, whilst birds were recorded as present on 19 further sites.

Pintail
Pintail occur as a rare breeding duck within the two districts. Between 11 and 41 pairs nested in Great Britain annually from 1974 to 1984, including up to four pairs breeding fairly regularly in north Caithness and occasional breeding records from north-west Sutherland (Thom 1986). Pintail were recorded as present on one survey site in south-central Sutherland.

Tufted duck
Tufted duck breed in small numbers in north-east Caithness and north-west Sutherland, but they tend to select more nutrient-enriched lowland waters. Additional birds visit the area in winter. Tufted duck were present on two sites, but there were no confirmed records of breeding.

Common scoter
The peatland lochs of Caithness and Sutherland form a stronghold for the common scoter within Britain (Figure 3.4e). The current British breeding population has been calculated as between 75 and 80 pairs, 30 of them within these two districts (Thom 1986; NCC and RSPB unpublished). An unpublished collation of breeding data by RSPB, however, suggests that there may be as many as 50 pairs breeding in Caithness and Sutherland. This evidence confirms the status of these peatlands as the most important breeding area for this duck in Britain. Assessment of precise numbers is made difficult by the secretive nesting of scoters in long heather some distance from open water, the disappearance of males early in the breeding season and annual fluctuations in the breeding population. Numbers appear to have reached a maximum in the late 1970s with a Scottish population of almost 100 pairs.

Common scoter

Common scoters were recorded from eight survey sites in Caithness and Sutherland. Between one and five pairs were seen at each of four sites in Caithness and three to eight pairs at each of four further sites in Sutherland. Whilst these numbers are small in absolute terms, the region holds a high proportion (about 40%) of the total British population of this species, and the protection of its nesting and feeding areas is a high conservation priority.

Goldeneye

Goldeneye are common winter visitors to, and passage migrants through, Caithness and Sutherland but have yet to be proved to breed. They are currently increasing their breeding range within Scotland (Dennis & Dow 1984), but breeding appears to depend on the availability of suitable nest-sites. Birds were present on 11 of the survey sites visited, but all these appeared to be wandering migrants or non-breeding birds.

Red-breasted merganser

This species has a boreal—low arctic distribution. Caithness and Sutherland hold both resident breeding and winter-visiting red-breasted mergansers. Although no population figures are available for the two districts, the species is widespread and thought to be currently increasing. Red-breasted mergansers were found on 19 of the survey sites. Three pairs were breeding on one site, one or two pairs bred on each of a further eight sites, and birds were present but with no evidence of breeding on 10 further sites.

Goosander

The goosander is a scarce breeding bird in the area and is probably resident throughout the year. Thom (1986) suggests that between 55 and 75 pairs breed in Ross and Sutherland. Both goosander and red-breasted merganser are subject to intense persecution in some areas of Caithness and Sutherland owing to their alleged depredations on fish. Goosanders were breeding on one survey site and birds were found to be present on five further sites, all in Sutherland.

3.6 Raptors
Hen harrier

Hen harriers are mainly resident in Caithness and Sutherland (Figure 3.4f), although there is a considerable dispersal of birds in the winter. Watson (1977) documented the persecution of hen harriers in Britain during the last century by gamekeepers and others. Available evidence suggests that the remote peatlands of Caithness and Sutherland were one of the last mainland strongholds for this species, and birds bred in this area until the turn of the century. The area was also one of the first to be recolonised, and Watson (1977) gives details of summer recoveries of ringed birds from Orkney found in Caithness and Sutherland during 1952-1970. By 1975, numbers had increased yet further in Sutherland but hen harriers were still sparse in Caithness.

Hen harrier

It appears that this recolonisation provided a focus for further expansion in mainland Scotland. Although numbers have now considerably increased from only a few years ago, the species is still persecuted, especially in areas of high grouse production. In the early 1980s there were thought to be 30-35 pairs in Sutherland, Caithness and Ross & Cromarty (Thom 1986), but more recent information suggests about 60 pairs in Caithness and Sutherland (RSPB unpublished data), though not all of these are strictly confined to peatlands.

Hen harriers were recorded as present on 16 of 63 survey sites (25%) but were proved to breed on only one site in Sutherland. Most sightings of hunting birds in summer are thought to relate to breeding birds whose territories overlapped survey sites, although their nests were outside them.

Sparrowhawk

Sparrowhawks have a very localised distribution in Caithness and Sutherland, and, whilst they are mainly resident, there is some dispersal out of, and influx into, the area in winter. Sparrowhawks tend to be associated with remnant native woodlands and also thicket-stage plantations, though they sometimes feed over nearby peatland. They are not so dependent on peatland habitats as other raptors in the region and were recorded on only one survey site (out of 50), in Sutherland.

Golden eagle

North-west Scotland provides one of the major breeding areas in Britain for golden eagles. The 1982 national survey identified 511 territories occupied by eagles (Dennis *et al.* 1984). There are thought to be about 50 pairs of eagle currently breeding in Caithness and Sutherland (R. H. Dennis pers. comm.), but some of these are coastal and thus cannot be considered to be dependent on peatlands (Figure 3.4g). Thirty territories contain significant areas of peatland habitat.

Eagles were noted as present on a quarter of the sites (15 out of 60) surveyed for them, but, as this species usually nests on crags, no breeding was recorded in this survey. It is clear, however, as other studies have indicated (Watson, Langslow & Rae 1987), that the blanket bogs are important feeding habitat for significant numbers of golden eagles.

Kestrel

There are no data on the size of the kestrel population in Caithness and Sutherland. Whilst a few birds are probably resident, the kestrel is largely a summer visitor here, but birds from elsewhere also visit the area in winter. Though more common in areas of marginal hill pasture, pairs of kestrels bred on two survey sites (out of 56) and were present on 13 further sites. It is clear that peatlands are of some significance as feeding areas for kestrels in Caithness and Sutherland.

Merlin

Merlins are small moorland falcons found throughout upland Britain but currently decreasing in numbers through much of their British range. The total British population size is about 600 pairs (Bibby & Nattrass 1986), of which at least 30 use the peatlands of Caithness and Sutherland (Figure 3.4h). Because of the difficulty in finding nesting merlins, this map undoubtedly under-records their breeding distribution. They probably occur more or less throughout the peatlands but at low density. They are partial migrants, and in autumn and winter merlins from Iceland also visit the region. Such information on breeding distribution as there is suggests that the species is commonest in west Sutherland, becoming scarcer further east into Caithness.

One pair of merlins was found on each of three sites (out of 52), with birds recorded as present on 13 further sites but not nesting within the area surveyed. Such a pattern of distribution suggests that merlins are widespread but nowhere common. Information collected by the MBS team in 1986 suggests that they are still subject to persecution in some areas of Sutherland.

Merlin

In a study of moorland birds in southern Scotland, Rankin & Taylor (1985) found a strong positive correlation between the breeding density of meadow pipits and that of merlins. In view of the feeding habitats of merlins, such a relationship is hardly surprising, but it emphasises the ecological links between one of the commonest and one of the scarcest peatland breeding birds.

Peregrine

Peregrines are widespread in Caithness and Sutherland, occurring both coastally and inland in both districts (Figure 3.4i). The population of this region constitutes less than 1% of the EC population of the peregrine, but most of the latter consists of the Mediterranean race *Falco peregrinus brookei*. The peatlands support a much greater proportion of the nominate race *F. p. peregrinus*. Some 35 pairs breed inland and depend on the peatlands for their feeding territories. Whilst nest sites occur on crags or in gorges, hunting takes place over a wide area and a great variety of prey, mainly birds, is taken (Ratcliffe 1980).

Out of 60 sites, a pair was recorded as breeding on one site in Sutherland and birds were present and seen hunting on 11 further sites. This undoubtedly under-records their use of the area, since they are easy to overlook owing to their tendency to fly very high when hunting.

Short-eared owl

Short-eared owls have a scattered breeding distribution, mainly in Caithness. The population varies to some extent with the availability of prey and thus can be greatly enlarged after 'vole years'. Though the species is resident, there is also considerable dispersal, mainly of young birds which leave the area where they hatch and settle elsewhere where there is abundant food. These wandering birds seem generally not to return to the breeding areas.

Out of 57 sites, one pair was recorded as breeding on a Sutherland site and birds were present on three further sites.

3.7 Other species
Red grouse

Red grouse are a widespread and resident game bird found throughout most of the area and belonging to the endemic race *Lagopus lagopus scoticus*, part of the widespread willow grouse species-complex. Whilst it is abundant

Red grouse

elsewhere in parts of Scotland, there is a special responsibility to safeguard this endemic race. Although widespread in Caithness and east Sutherland, grouse tend to be more localised in distribution in west Sutherland, perhaps owing to the greater fragmentation of their moorland habitats. There is little good information on population size, but birds were found on all survey sites and presumably bred on all. Earlier in this century there were a number of productive grouse moors in the area, especially in Caithness, but, as elsewhere, there has been a long-term decline in numbers through much of the two districts. The causes are still being studied by the Game Conservancy elsewhere in Scotland. Caithness has also been a traditional area for the sport of 'grouse-hawking' with trained falcons, for the large, flat moorlands make it easy to follow flights of quarry over long distances and give the best chance of recovering the valuable hawks.

Black grouse

Black grouse are resident in Caithness and Sutherland. They are rare in Caithness but locally more frequent in Sutherland, and there has a been a slight extension in their range as the edges of new conifer plantations are colonised. Moss (1986) has recently reviewed the distribution of black grouse in Scotland and has shown that the species is ecologically more tolerant of high summer rainfall than other game birds such as capercaillie. This tolerance of high rainfall allows exploitation of peatland areas in the wetter north and west of Scotland.

Birds were recorded as present on only two (of 67) sites, and the species cannot thus be regarded as an important component of the peatland bird fauna in Caithness and Sutherland.

Arctic skua

Arctic skuas, with a boreal—high arctic distribution, have long bred on the peatlands (Everett 1982) and, although numbers have never been large, are a significant component of the moorland bird fauna. Yeates (1948) drew attention to the occurrence of small colonies of breeding arctic skuas, particularly in Caithness. There has been little documentation of numbers, however, until fairly recently. Everett (1982) recorded 20 pairs in 1969-70 and 28 pairs in 1974 in thorough surveys of the whole of Caithness. These totals are markedly less than the minimum population of 40 pairs found by Reed, Langslow & Symonds (1983b) on UBS survey sites in 1979 and 1980. When the UBS surveys were compared with previous information, it was found that there were at least six new breeding areas in 1979–80 compared with 1974 and, of the six sites surveyed in 1979–80 which had held birds in 1974, five held larger numbers — a significant increase (Reed, Langslow & Symonds 1983b).

Arctic skua

These results suggest that arctic skuas are still increasing their range within Caithness, although there is no more recent comprehensive information than that of 1979–80. There still appears to be suitable habitat unoccupied within Sutherland; although Thom (1986) recorded one to three pairs breeding on peatlands there, none of the UBS/MBS plots in that district contained skuas.

Black-headed gull

Black-headed gulls are breeding summer visitors to the area, but passage migrants and winter visitors also occur. The species is common and widespread, although there is no good information available on total population size for the two districts. Black-headed gulls breed mainly on water-bodies where there is floating vegetation or where there are islands separated from the land. Thus breeding sites are somewhat irregularly distributed, depending on the physical nature of particular lochs.

A total of 15 pairs was recorded

breeding on one site in Sutherland and four pairs bred on a site in Caithness. Birds were present at 16 other sites throughout the peatlands.

Common gull

Common gulls breed throughout Caithness and Sutherland; they are mainly summer visitors, but some may be resident throughout the year. Elsewhere they have a northern continental—low arctic distribution. There are no good data on the overall population size, but, on the basis of the breeding distribution shown in *The atlas of breeding birds in Britain and Ireland* (Sharrock 1976) and average colony size, some 4,000 (or 10% of the British population) probably breed within Caithness and Sutherland. Breeding takes place in a variety of habitats, including moorland, bog, islets, rocky shores of lochs and shingle banks alongside larger water-bodies. There is thus abundant suitable habitat within the peatlands.

Common gulls were recorded breeding on six (of 59) sites, which held between two and 20 pairs each. Birds were present at 16 further sites, and, whilst there was no evidence of breeding, they were probably nesting just outside the surveyed sites.

Lesser black-backed gull

Lesser black-backed gulls are breeding summer visitors to Caithness and Sutherland, though some also occur as passage migrants. Their breeding distribution is mainly coastal, but some also occur inland. The species was recorded on 11 (of 58) survey sites, but there was no evidence of breeding at any of these. Former breeding colonies of this gull, with smaller numbers of the next two species, were mentioned as occurring in the flow country by Harvie-Brown & Buckley (1887). They were said to be heavily persecuted and have evidently declined considerably since that time.

Herring gull and great black-backed gull

Herring gulls are a common resident species in Caithness and Sutherland, but their breeding distribution is mainly coastal. Birds were recorded from four (of 58) sites, but these were presumably either migrants, non-breeding wanderers or visitors from the coast.

Like herring gulls, great black-backed gulls have a coastal breeding distribution and, although they were recorded from 18 (of 58) sites, there was no evidence of breeding at any of these. The status of birds seen was probably similar to that of herring gulls observed inland.

Common and arctic terns

Both these species have a predominantly coastal breeding distribution, though there are some inland colonies of common terns, usually associated with the larger lochs. Amongst these colonies Laybourne, Manson & Collett (1977) have found arctic terns nesting, and this species has also been found at a small dubh lochan within quaking *Sphagnum* bog. The lochs were some 16-20 km from the coast, and arctic terns were not seen to feed inland (Laybourne, Manson & Collett 1977). Common terns were seen during NCC surveys at one site, and arctic terns were seen on two sites. In neither case was there any evidence of breeding.

Cuckoo

As elsewhere, cuckoos are breeding summer visitors to Caithness and Sutherland, particularly parasitising the nests of meadow pipits. There is little available information on numbers, although the species occurs widely through the peatlands. It was recorded as present on only two sites, but the survey techniques used are unlikely to have reflected accurately the true abundance of this species.

Skylark

Skylarks breed commonly throughout the peatlands of Caithness and Sutherland and, after meadow pipits, are the commonest breeding passerine on these blanket bogs. Although a few birds are probably resident, the species is mainly a summer visitor to the north of Scotland.

Skylarks were recorded breeding on all the survey sites. However, because of the abundance of this species, it has always been difficult to estimate breeding densities. Counts vary from an average of 210 birds recorded on visits to a site in Sutherland to only one singing male recorded from another site in the extreme north-west of Sutherland. Skylarks tend to be commoner in more grass-dominated areas which occur over shallow peat or mineralised soils.

Meadow pipit

The meadow pipit is a common and widespread breeding bird throughout the peatlands of Caithness and Sutherland and is probably the commonest breeding passerine. Numbers of passage migrants and also some winter visitors use the region, whilst the breeding birds winter mainly in the Iberian peninsula. The small number of birds that winter in Caithness and Sutherland, mainly on the coast, are probably immigrants rather than locally breeding birds.

Meadow pipits were found breeding on all sites surveyed, but, as with skylarks, precise breeding numbers have been hard to estimate. The highest average count was of 249 per visit to a site in Sutherland. Greatest numbers occur in heather-dominated areas, in contrast to the skylark.

Grey and pied wagtails

Both grey and pied wagtails were found on the sites surveyed. Grey wagtails are breeding summer visitors to the peatlands, wintering in England, Ireland and France. Whilst fairly widespread in south-east Sutherland, they are scarcer elsewhere on the blanket bogs of the two districts. Out of 60 sites, a pair was found breeding on one site in Caithness, whilst birds were present on two other sites. Many of the watershed mire systems, with little water movement, are probably unsuitable for this species, which requires fast-flowing water.

Pied wagtails are more common than grey wagtails, being widespread in both Caithness and Sutherland. Out of 57 sites, one or two pairs were found on each of 10 sites and birds were present on 17 further sites. White wagtails are passage migrants through the area in considerable numbers. Occasional pairs breed, usually hybridising with pied wagtails.

Dipper

Dippers are resident water birds with a widespread distribution in Caithness and Sutherland determined by the presence of suitable streams for breeding and feeding. Numbers are unknown, but one or two pairs were found on each of five sites (out of 57) and birds were present on 12 further sites.

In mid-Wales, the population of dippers has been the subject of intense autecological study (Ormerod 1985; Ormerod, Boilstone & Tyler 1985; Ormerod, Tyler & Lewis 1985; Tyler & Ormerod 1985). In particular, the factors affecting distribution and abundance have been closely investigated. In a detailed study of the influence of stream acidity on breeding distribution, Ormerod, Tyler & Lewis (1985) have shown that breeding density is reduced on those stretches of river which have raised acidity. This is probably caused by the effect that raised acidity has in lowering the abundance of freshwater invertebrate prey. In areas of highly acid rocks and soils, which have low buffering capacity, coniferous afforestation has been shown to increase the acidity of run-off. Ormerod, Tyler & Lewis (1985) stated that all the streams which showed evidence of an historical fall in pH drained from catchments which were 25-40% covered by mature forests; none had breeding dippers within 8 km of its source in 1982.

The Welsh studies suggest that, if coniferous afforestation of the peatlands of Caithness and Sutherland resulted in significant increase in stream acidity, dippers would be one of the first water birds to be seriously affected.

3

Other passerines

Whilst not commonly thought of as birds of peatlands, wrens were found on 11% of 54 survey sites. Wrens are resident in Caithness and Sutherland, though the severity of some winters, to which the species is sensitive, means that numbers are never great. One to three pairs were found nesting on each of five sites and birds were present on a further site.

Whinchats are breeding summer visitors in the peatlands and are widespread, although scarcer in Caithness than in Sutherland. Out of 60 sites, three pairs were recorded nesting on one site in Sutherland and birds were present on two further sites.

Wheatears occur in Caithness and Sutherland both as breeding summer visitors and as passage migrants from Iceland and Greenland. They are widespread and common on the peatlands, usually nesting where there are outcrops of rocks and dry grassland. One to seven pairs nested on each of 17 (out of 41) sites and birds were present on 19 further sites.

Ring ouzels are widespread in most of Sutherland, but they are scarcer in easternmost Sutherland and Caithness. Five pairs nested on one site, one pair on another, and birds were present on two further survey sites out of a total of 57. The species requires steeply sloping ground and was thus absent from most of the sites surveyed for waders, particularly those dominated by watershed mire formations.

Sedge warblers are widespread in suitable habitat in Caithness, though scarcer in Sutherland. A pair was found nesting at one site and birds were present at five further sites out of 54.

After meadow pipits and skylarks, hooded crows were the most commonly recorded passerines on survey plots, being seen on 70% of 57 sites. Single pairs bred on three sites and birds were present on 37 others. The species is resident and widespread throughout the two districts, with hybrid hooded/carrion crows occurring locally.

A single pair of ravens bred on one site, and birds were recorded during surveys on 20 further sites out of 57. The species is common and widespread, especially in west Sutherland, though scarcer in Caithness.

Reed buntings are widespread in Caithness but scarcer in Sutherland. One or two pairs were found breeding on each of two (of 67) sites.

3.8 Conclusions

The sample surveys, covering about one-fifth of the Caithness and Sutherland peatlands, have established that the region contains particularly large and diverse moorland breeding bird populations. While the breeding densities of some species such as golden plover, curlew, red grouse and merlin are appreciably lower than in some moorland areas further south, the total species list for these peatlands with their associated open water habitats greatly exceeds that for typical southern moorlands. The huge extent of these peatlands also results in large total populations for many species.

Earlier records of breeding birds for the region are mostly unquantified, so that comparisons are hardly appropriate. The subjective impression, however, is that these peatlands have changed little, if at all, in their most notable ornithological features since this interest was first discovered over a century ago. Some of the inland nesting colonies of gulls have certainly decreased or disappeared, and golden plovers may no longer reach the high densities reported from Strathmore around 1920. Certain species may have had minor fluctuations but retained an overall status hardly differing from one decade to the next. The few rarities may be new colonists responding to more favourable climate since 1955, or they may simply have been found through the much greater ornithological interest in the region in recent years.

It therefore seems quite safe to regard the recent advent of large-scale afforestation as a quite unprecedented environmental change in its potential for impact on these peatland bird populations.

Figure 3.3
Breeding densities (pairs/km²) of wader species on 77 peatland sites in Caithness and Sutherland.

In each histogram the left-hand column shows the number of sites on which the species was not recorded as breeding.

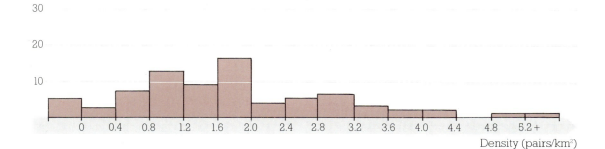

Frequency Figure 3.3a Golden plover

Frequency Figure 3.3b Dunlin

Frequency Figure 3.3c Greenshank

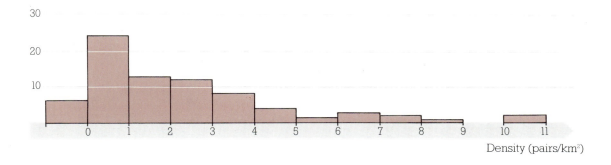

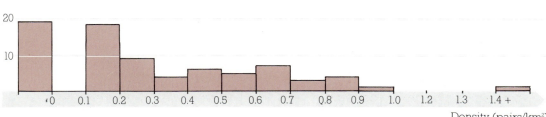

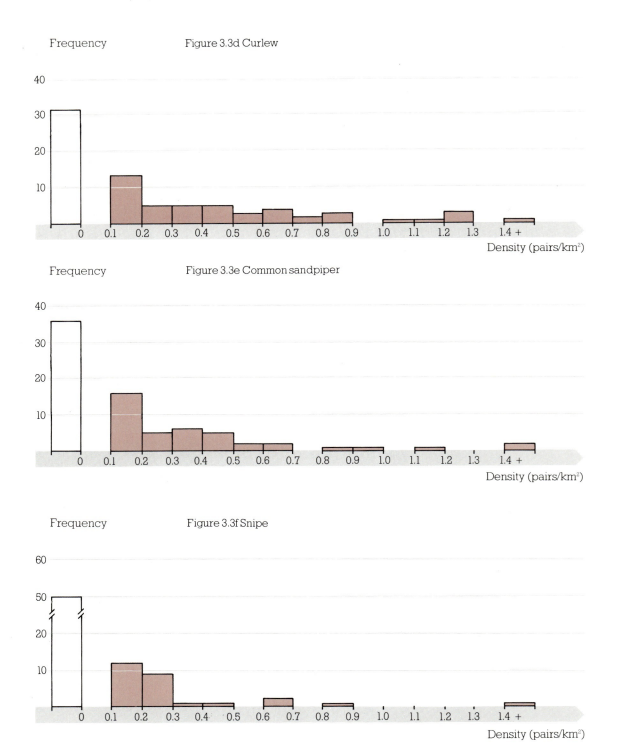

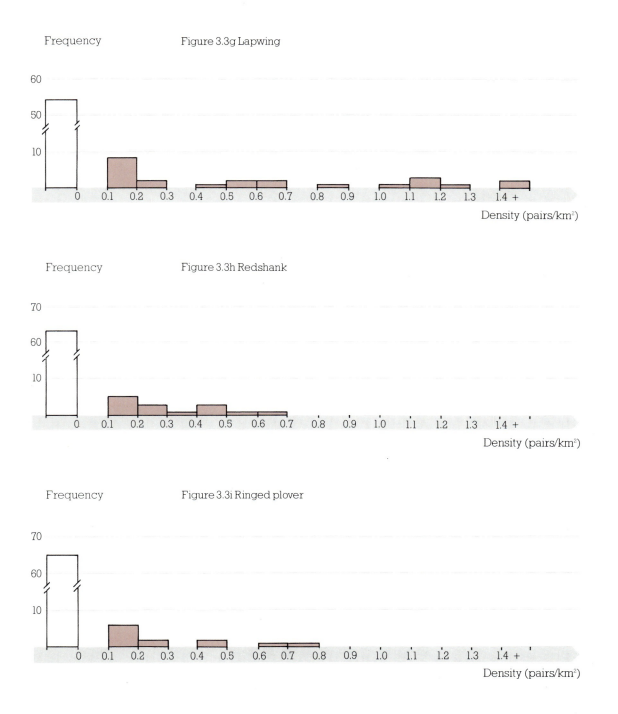

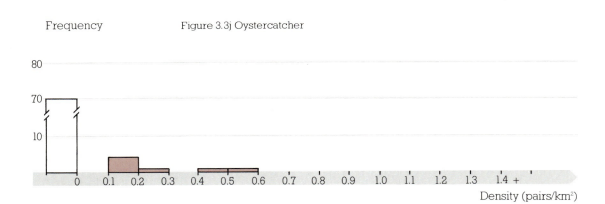

Figure 3.3j Oystercatcher

Figure 3.4
Breeding distributions of nine bird species in 10km grid squares throughout Caithness and Sutherland.

From Sharrock (1976), with additional information for the three wader species from the surveys presented in this report.

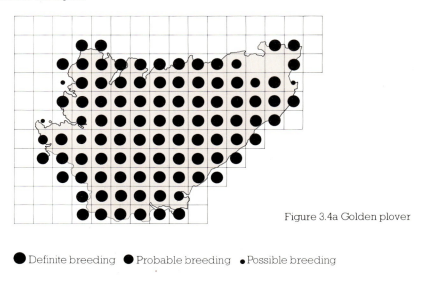

Figure 3.4a Golden plover

● Definite breeding ● Probable breeding · Possible breeding

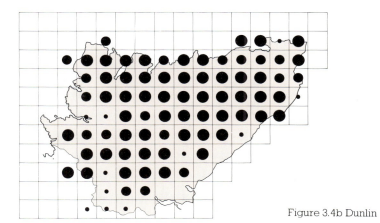

Figure 3.4b Dunlin

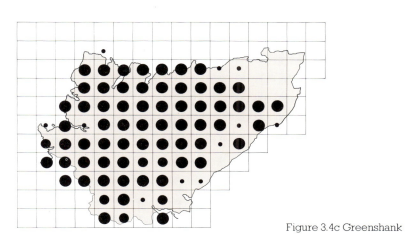

Figure 3.4c Greenshank

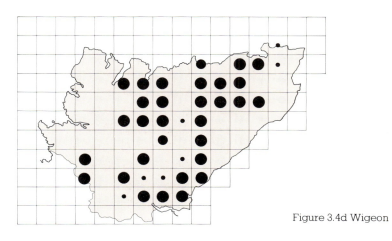

Figure 3.4d Wigeon

● Definite breeding   ● Probable breeding   • Possible breeding

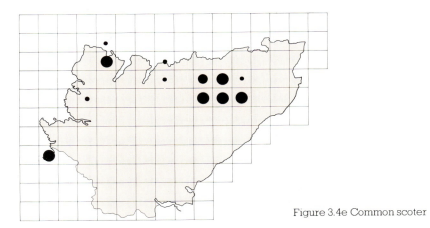

Figure 3.4e Common scoter

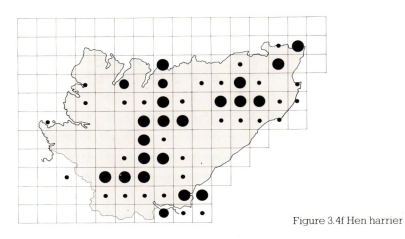

Figure 3.4f Hen harrier

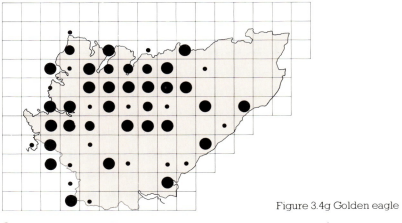

Figure 3.4g Golden eagle

● Definite breeding   ● Probable breeding   · Possible breeding

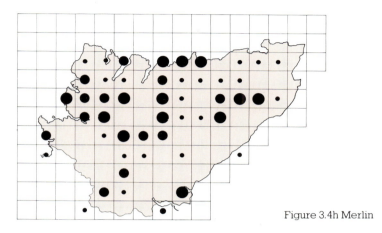

Figure 3.4h Merlin

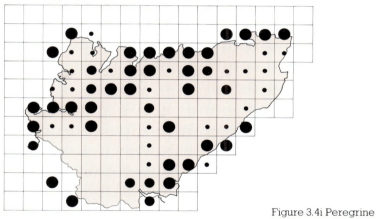

Figure 3.4i Peregrine

● Definite breeding ● Probable breeding • Possible breeding

Overall distribution and numbers of peatland birds in Caithness and Sutherland

4.1 Introduction

Only about one-fifth of the peatlands of the region have been surveyed in detail by the standard method described in the Appendix. Yet the 77 plots surveyed, covering 51,929 ha (Table 4.1: 19% of the remaining area suitable for breeding waders), required 433 separate visits during 22 man-summers of fieldwork over the period 1979-1986. There are obviously considerable logistical and financial constraints to surveying the remaining 81% of the peatlands to the same standard. Moreover, when afforestation is occurring so rapidly and on such a wide and haphazard geographical scale, we cannot afford to await such completion and must find a more rapid yet reliable means of assessing the unsurveyed areas.

This chapter accordingly has four aims:
- to find out if there are consistent relationships between breeding bird densities and habitat features recognisable on standard Ordnance Survey maps;
- to use any such associations, in combination with information on such habitat features derived from the maps, to predict the ornithological quality of the unsurveyed areas;
- to examine the further possibility of using such associations between breeding densities and map attributes to estimate the total bird populations found in each habitat category and in the total peatland area;
- to enable pre-afforestation maps to be used to assess the previous ornithological quality of land now afforested and, from this, to estimate the reductions in populations, on the same principle as that used for estimating the present populations.

4.2 Wader densities and habitat characteristics

Background considerations

The possibility of making such interpolations will necessarily be limited to those species which have a virtually continuous distribution over the whole peatland area and are sufficiently numerous. There is also obvious merit in applying such an approach especially to species with high conservation value. In practice the study has been limited at present to the three most abundant, characteristic and important wading birds of the peatlands, the golden plover, dunlin

Table 4.1 Land-use and ornithological survey in Caithness and Sutherland

Category	Area (ha)	Category as percentage of land area
Total area of Caithness and Sutherland	764,094	100%
Ancient, semi-natural and long-established woodland	12,204	1.60%
Forestry plantations	73,046	9.56%
'Improved' agricultural land and human settlements	104,090	13.62%
Fresh water (minimum area)	25,170	3.29%
Land too steep/high for moorland waders (including some high-altitude blanket bog) and coastal areas	279,484	36.58%
Remaining area of blanket bog currently suitable as breeding habitat for moorland waders*	270,100	35.35%
Total moorland surveyed by UBS/MBS (included in last category)	51,929	6.80%

* The area of peatland recorded as suitable habitat for breeding waders does not equate to the full extent of blanket bog. There are considerable areas of steep, high-altitude blanket bog that are not considered suitable breeding habitat.

and greenshank.

Reed (1985) and Reed & Langslow (in press) have previously used Upland Bird Survey data to correlate breeding densities of waders with certain landform characteristics identifiable from 1:25,000 Ordnance Survey maps (Table 4.2). Using these statistically derived relationships, it is possible to assess the likely densities of waders on unsurveyed sites by measuring the extent of these landform features from maps. For example, golden plovers, dunlins and greenshanks were all found to prefer 'flow' bog with pool complexes. High greenshank numbers are usually found where these pools are widely but regularly spaced over large areas. Drier and steep moorlands are poorly used by all species, and sites close to conifer plantations hold lower numbers than would otherwise be expected (Stroud & Reed 1986).

The 1:25,000 Ordnance Survey maps, however, show only certain physical features of these peatlands (e.g. dubh lochans, pool complexes and probable wetness as reflected by contouring). Waders also respond to vegetational features (Table 4.2), many of which are closely determined by physical structure, so that, for example, the presence of bog pools determines the presence of associated pool-side plant communities. However, other vegetation is not linked with physical features in this way: a site otherwise expected to have a favoured mix of habitats may be too frequently burnt, heavily grazed or otherwise modified. Such operations can markedly influence the vegetation, yet are not apparent until the site is actually visited.

Thus, although assessments of peatland quality for waders from map evidence alone are possible if the habitat requirements of the species concerned are known and if the

Table 4.2 Habitat preferences of breeding waders on moorlands in Sutherland
Updated from Reed & Langslow (in press). See section 3.4 for explanation of greenshank habitat use. Habitat preferences encompass all the summer activities during the breeding season and include habitats selected for their feeding, nesting and young-rearing potential. The surveys did not aim to locate nests; they have tended to record birds in their most visible locations, often at or near feeding places.

	Vegetation type							Vegetation height			Vegetation age			Wetness			
	Trichophorum/ Myrica mire	Calluna/ Eriophorum mire	Juncus flushes	Dry Trichophorum	Dry Calluna sward	Grass patches	Mosaic of vegetation types	<10cm	10-20cm	>20cm	Young	Medium	Old	Pool/dubh lochan complexes	Bog	Damp	Dry
Golden plover	+	+	●	—	—	●	●	+	+	+	+	●	●	+	+	+	●
Dunlin	+	+	●	—	—	—	●	+	●	—	+	●	●	+	+	●	—
Greenshank	+	●	—	—	●	—	+	+	●	—	+	+	●	+	+	+	+
Curlew	—	—	+	—	—	●	+	+	+	+	+	●	—	—	—	+	●
Snipe	—	—	+	—	—	+	+	+	+	+	—	—	—	●	—	+	—
Redshank	—	—	+	—	—	+	+	+	+	+	+	●	—	—	—	+	—

Key: + preferred
— avoided
● no obvious trend

associations between habitats and physical features shown on maps are understood (Campbell 1985), there are risks that such assessments may be faulty because of various management practices which degrade otherwise 'good' sites.

An attempt was made to check the accuracy of the assessments, on which depends any attempt to interpolate the results over the rest of the Caithness and Sutherland peatlands. This was done in three stages. First, a preliminary investigation was undertaken to establish whether map information could be used to predict densities of waders with an acceptable degree of reliability. Secondly, the habitat features shown on maps and related to different densities of waders were categorised more formally. Thirdly, a direct test was undertaken.

Preliminary investigation

Each of the 24 sites surveyed by NCC from 1982 to 1984 was allocated by a worker who had not been involved in the surveys to one of five quality classes, from expected high breeding densities of waders to expected low densities. This was done separately for golden plover, dunlin and greenshank on the basis of the physical characteristics shown on maps of each site and thus the predicted vegetation (see Table 4.2). The expected density scores were then compared with the actual densities found at each site. (For sites surveyed in more than one year, the mean observed density was used.)

Figure 4.1 shows the relationship between densities observed at these sites for golden plover, dunlin and greenshank and the estimations of site quality (in terms of expected density) for each species. An analysis of variance was undertaken to see whether there were significant differences between the mean densities in each of the five quality classes.

The differences between the densities observed in the five quality classes were greatest for greenshanks ($F_{4,19} = 44.19$; $p \ll 0.001$). This is because greenshank densities were found to be strongly related to the distribution of lochs and dubh lochans within blanket bog and because these features are especially easy to recognise from maps. Mean densities of dunlins also differed between classes of sites ($F_{4,19} = 6.28$; $p < 0.01$), and this was also the case for golden plovers ($F_{4,19} = 5.79$; $p < 0.01$).

Much of the variation in densities between sites assigned to the same quality class could be attributed to habitat degradation, such as overgrazing and burning, which was not identifiable from maps. However, it was clear that for each of the three main peatland wader species the areas with the highest densities could generally be identified from maps, and the method was considered reliable enough to develop further.

Categorisation of habitat types

Based on knowledge of peatlands and the results of the preliminary investigation, four categories of peatland landform were defined more formally. The features used in formal identification of categories from maps are given in Table 4.3, and the following descriptions amplify these in relation to what can be seen on the ground.

Category A comprises the very wettest areas of peat, with numerous pool complexes and extensive *Sphagnum*-dominated flows set in a bog-covered landscape. Maps show high densities of dubh lochans clustered into obvious pool complexes but there are also scattered larger lochans. The ground is either virtually flat or gently sloping, with no steep gradients. There are few, if any, rocky outcrops and no crags are shown on maps.

Category B consists of sloping blanket bog with pools and is drier. There is a low density of pools, set more or less discretely on gentle slopes, and the ground consists of gentle ridges and watersheds but is not flat. Rocky outcrops and drier morainic features, whilst not numerous, can be evenly spaced across wide areas. There are often numerous larger lochs within the landscape. The blanket bog communities are drier and less

4

Category A: Pool complexes and wet *Sphagnum* flows. These areas have a characteristically high density of bog pools which are attractive to breeding waders and waterfowl. Loch Syre area, Sutherland, July 1986

Category B: Sloping blanket bog with individual pools. This drier type of blanket bog covers large areas of Caithness and Sutherland. Ben Hutig area, Sutherland, April 1986

Category C: Steeper and more broken ground. This landform is often broken by underlying bedrock protruding from the blanket peat and has generally low densities of breeding waders. Inchnadamph National Nature Reserve, Sutherland, July 1983

Category D: Montane and other unsuitable areas. Fell-fields such as these are unsuitable as habitat for moorland breeding waders

Table 4.3 Features used in categorisation of the Caithness and Sutherland peatlands into four landforms for estimation of breeding wader populations. Landforms were assessed from 1:25,000 maps

Category	Dubh lochans	Streams	Lochs	Topography	Gradients
A: Pool complexes and wet *Sphagnum* flows	High density of pools and dubh lochans clustered in obvious pool complexes. Usually at least one pool complex or marsh symbol per 1 km². Pool complexes with more than 10 pools per complex.	Few streams, with none issuing from watershed mire pool complexes.	Scattered larger lochs, usually with gently curved edges indicating peaty banks.	Flat, open and obviously boggy.	Gently sloping to flat, with very low gradients. Generally less than five 25-ft contours crossed per 1 km² diagonal at 1:25,000.
B: Sloping blanket bog with pools	Low density of pools set more or less discretely. Pools less than 10 per complex, or less than one complex per 1 km².	More streams, often branched into dendritic drainage systems hillsides.	Larger lochs often numerous and irregular in shape.	Gentle ridges and watersheds.	Generally low gradients, but sloping gently — not flat. Usually from five to 12 25-ft contours crossed per 1 km² diagonal at 1:25,000.
C: Steeper and broken ground	No marsh symbols or dubh lochans marked on 1:25,000 map. Very few, if any pools.	Streams and waterfalls down steep slopes.	Few large lochs, although sometimes surrounded by steep banks.	Hillsides and broken or rough ground indicated on map.	Gradients steeper: more than 13 25-ft contours per 1 km² diagonal at 1:25,000.
D: Montane and other unsuitable areas	None.	Linear, with many waterfalls and streams descending steep slopes.	Small lochs usually in corries, often steeply embanked.	Mountainous, with considerable areas of bare rock/scree shown on map.	Very steep slope, usually more than 25 25-ft contours per 1 km² diagonal at 1:25,000.

Sphagnum-dominated than in the previous category, and the surface vegetation is often eroded into gullies, either naturally or by overgrazing or after severe fires, exposing areas of bare peat, which may be extensive.

Category C is that of steeper and more broken ground. Here the gradients are steeper, with no substantial areas of pools or dubh lochans. The ground is often highly eroded, with few, if any, wet *Sphagnum*-dominated areas. Rocky outcrops and dry morainic features are abundant. Podsolic and gley soils with shallow peat surface horizons prevail, rather than true blanket peat. The vegetation tends to have *Trichophorum cespitosum*, *Molinia caerulea* or *Calluna vulgaris* dominant and is of the type characterised as 'wet heath' or, in the driest situations, acidic dwarf shrub heath.

Category D is the steepest ground, with screes, outcrops, crags, high montane watersheds and summits with fell-field and shallow montane blanket bog. This was considered to be unsuitable habitat for most moorland waders. Because of the very low wader densities found during preliminary work in these steep moorland and rocky areas, sample plots generally excluded such areas, and it does not form part of the analysis.

A test of the landform/wader associations

Sites surveyed by the NCC and RSPB teams were randomly assigned to one of two groups. One group of sites (the control set) was examined by someone familiar with most of the sites and the survey results. Each site was divided into landform categories according to Table 4.3, and the area of each category within each site measured to the nearest 1 ha. A very few sites (four out of 38) showed such diversity of landforms in a small area that such division was not possible and these were not considered further.

The largest category of landform on each site was then singled out and the numbers of golden plovers, dunlins and greenshanks that had been found by survey teams within that area were assigned to it. These numbers were then expressed as densities for each species within that landform category. In a few instances, where a site contained roughly equal areas of two landforms, it was possible to calculate densities for both landform types on the same survey site.

Figure 4.2 shows the distributions of densities of the three main wader species with respect to landform categories A, B and C. After densities had been obtained from the control set of sites, the data were logarithmically transformed ($\log(x+1)$), in view of the skewed nature of the density distributions (see section 3.4), for calculation of means and standard deviations, which are given back-transformed.

The categorisation of landforms provided overlapping but clearly different distributions of wader densities, as indicated below.

Golden plover
Of the control sites where landform A was recorded, 73% contained densities greater than $1 \cdot 6$ pairs/km^2. In the case of landform B, 54% of areas fell within the range of $0 \cdot 8 - 1 \cdot 59$ pairs/km^2, with equal numbers above and below these limits. Category C areas had the lowest densities, with 70% of such areas holding less than $0 \cdot 8$ pairs/km^2.

Dunlin
Dunlins showed clear separation into three groups of differing densities. In category A, 83% of areas held dunlin densities in excess of $1 \cdot 0$ pairs/km^2. These high densities reflect the strong affinity of dunlins for dubh lochans and other wet areas. In category B, 38% of dunlin densities fell in the range $0 \cdot 2 - 0 \cdot 99$ pairs/km^2, whilst in category C 90% of the densities fell below $0 \cdot 2$ pairs/km^2; indeed, eight out of the 10 areas in this last category held no dunlins at all.

Greenshank
The categorisation of greenshank densities was less clear than for dunlins or golden plovers. Category A areas were quite clearly defined, however, with 50% of them having densities greater than $0 \cdot 3$ pairs/km^2. In category B, 23% of areas held densities between $0 \cdot 1$ and $0 \cdot 29$ pairs/km^2, but a large number of areas (nine out of 13) held no greenshanks at all. In category C, 90% of areas held none. Because of the similarity in the distribution of greenshank densities in categories B and C, these two categories were combined for this species. Whereas 78% of B and C areas combined held no greenshanks, only 28% of A areas held none. Greenshanks show more complicated habitat selection than the previous two species (section 3.4): they need running or standing open water for feeding, but they can nest at some distance, either on hummock bog or on drier and often morainic ground on moderate slopes.

If information from maps provides a sound basis for estimating densities of breeding waders, classification of a second sample of sites according to the landform features in Table 4.3 should result in similar distributions of breeding densities. To investigate this, the second group of randomly assigned sites (the test set) was assessed, using only Ordnance Survey maps and the diagnostic classification in Table 4.3, by a worker unfamiliar with birds or bird surveys but familiar with peatland habitats. This worker classified the landforms within each site and marked their limits on the map. The area of each landform category in each site was then measured and the numbers of birds which had been found in each area by field survey assigned to it. (The last set of information was not available to the worker undertaking the classification.) The distribution of densities was then calculated as for the control set of data.

If the map information is a useful means of interpolating bird densities to unsurveyed sites within the region, mean densities in each category should not differ significantly between control and test sets. This proved true for all the comparisons (Table 4.4). Variances were also generally similar between the two sets. The only exceptions were in category C for golden plovers and in category A for greenshanks, for both of which the control set was slightly more varied than the test set. The use of some eroded areas by golden plovers and the complex habitat selection of greenshanks were the probable

4

Figure 4.1 Relationships between predicted and observed densities of three wader species on 24 sites in Sutherland (1982-1984)

Figure 4.1a Golden plover

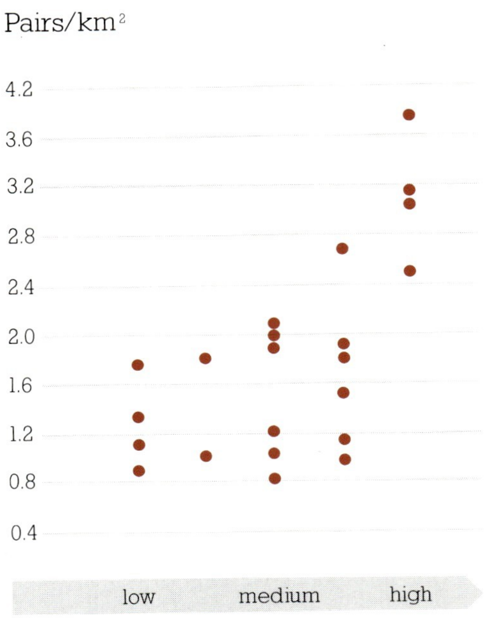

Pairs/km²

Figure 4.1b Dunlin

Pairs/km²

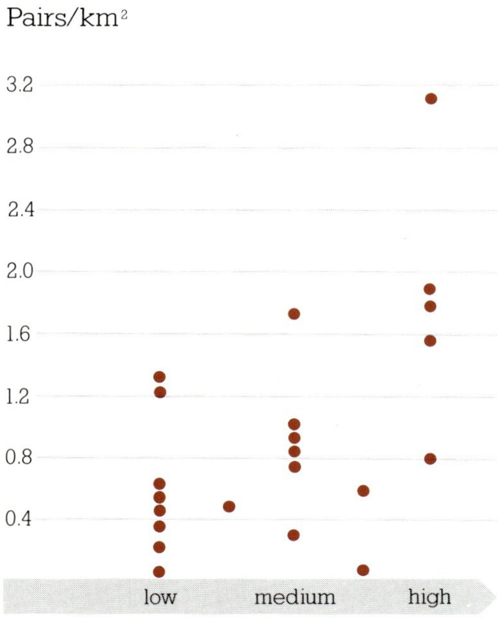

Figure 4.1c Greenshank

Pairs/km²

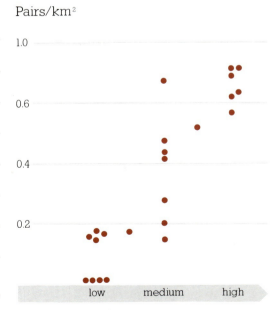

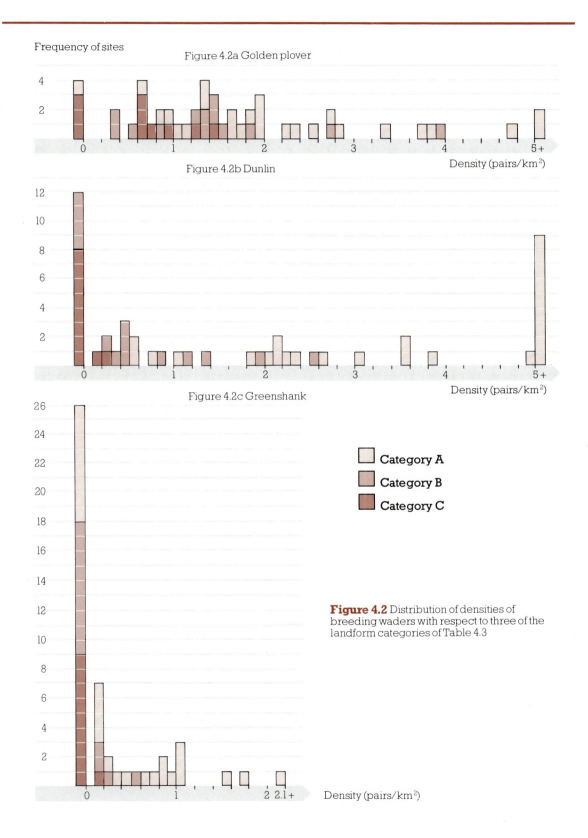

Figure 4.2 Distribution of densities of breeding waders with respect to three of the landform categories of Table 4.3

causes.

The correspondence of density distributions between the control and test sets demonstrates that:
- the categories of peatland (Table 4.3) are real and reflect identifiable habitat types, and
- it is possible to identify these categories on the basis of map evidence alone.

This has important implications, since it means that the quality of bird habitats of the whole area of blanket bog of Caithness and Sutherland outside survey sites can be assessed, as in the following sections.

4.3 The identification and mapping of areas of peatland suitable for breeding waders

Introduction

In order to estimate the breeding wader population of the two districts, a mapping exercise was carried out, first to measure the area of the main landforms and land-uses, secondly to assess the quality of the peatlands as a habitat for breeding waders, and finally to estimate the distribution and abundance of the three main species of waders occurring on these blanket bogs.

A set of 1:25,000 maps of the two districts was marked with the position and extent of several land-use types — forestry, 'improved' agricultural land and human settlements — based on information available in August 1985. Additionally, all major water-bodies were marked. Limited survey of steep ground and high peatlands had previously shown that they contained very low densities of breeding waders (see section 4.2). These could be classed as unsuitable on the basis of their physical characteristics, and such areas also were indicated on the maps. It was assumed that the remaining areas held breeding waders on the basis of associations of wader distribution and mapped features. Many of the categorisations made in this mapping exercise were checked on the ground in 1986. Land-use categorisation from map evidence was found to be largely accurate in areas checked on the ground. After the maps had been annotated, the area of each land-use type within the two districts was calculated. The extent of each type was

Table 4.4 Comparisons between control and test data-sets as to means and variances in an investigation of the categorisation of landform types
Data were transformed by log (x + 1) on both F and t tests. For greenshank, categories B and C have been combined for reasons given in the text.

	Category A	Category B	Category C
Golden plover	$F_{25,18} = 2.128$ $p>0.05$	$F_{5,12} = 2.342$ $p>0.05$	$F_{9,4} = 13.269$ $0.01<p<0.05$
	$t_{43} = 0.219$ $p>0.05$	$t_{17} = 0.744$ $p>0.05$	$t_{13} = 0.660$ $p>0.05$
Dunlin	$F_{18,25} = 1.215$ $p>0.05$	$F_{5,12} = 1.083$ $p>0.05$	$F_{4,9} = 2.760$ $p>0.05$
	$t_{43} = 0.283$ $p>0.05$	$t_{17} = 0.368$ $p>0.05$	$t_{13} = 0.136$ $p>0.05$
Greenshank	$F_{25,18} = 3.074$ $0.01<p<0.05$	$F_{10,22} = 1.655$ $p>0.05$	
	$t_{43} = 0.245$ $p>0.05$	$t_{32} = 0.692$ $p>0.05$	

estimated in units of 10 ha. A similar method was used by Angus (1987) when calculating peatland extent in the same region.

Details of forest extent and ownership were updated to January 1986 from information made available by the Forestry Commission. All woodland was measured. In addition, NCC's inventories of ancient, semi-natural and long-established woodland in Caithness and Sutherland (Walker 1985, 1986), which distinguish between these and more recent plantings, were consulted.

Results — current extent of habitats in Caithness and Sutherland

The total mapped area for both districts came to 792,390 ha. This compares with local government figures of 586,518 ha for Sutherland and 177,576 ha for Caithness, totalling 764,094 ha (Table 4.1). Thus the mapping discrepancy is 3·7%. Most of this occurred in the delineation of boundaries in the intertidal zone. This is one of the types of land unsuitable for breeding waders, so the total for such land has been corrected accordingly. The natural habitats unsuitable for moorland waders are unevenly distributed across the districts, being found largely in the mountainous west of Sutherland. The total extent of blanket bog is greater than that of moorland suitable for breeding waders. The rest of the peatland complex nevertheless contains other habitats and species of conservation importance.

The area of 'improved' agricultural land and settlements is 104,090 ha. This is a broad category, including areas of improved hill pasture as well as better arable land in Caithness. It is concentrated in north-eastern Caithness, with marginal crofting land on the coasts and in some of the larger Sutherland straths.

Within Caithness and Sutherland at least 79,350 ha were either already planted or programmed for afforestation by January 1986. The total extent is slightly greater, as several small areas on some estates and farms have been planted as shelter-belts and there have also been plantings close to roads to act as snow barriers and windbreaks. The areas afforestable in the short and medium terms further increase the total of 79,350 ha, which does not include a considerable area of land owned by forestry interests for which grant-aid has not yet been sought.

Distribution of woodland

In Caithness there is little semi-natural woodland (Walker 1986). The total area of ancient, semi-natural woods and long-established plantations amounts to only 677 ha in the whole district. Most of these woods are found in sheltered glens or coastal gorges. Sutherland is more naturally wooded, with a total of 11,527 ha of ancient, semi-natural and long-established woodland. In west and central Sutherland there are many small, semi-natural woodlands, and there are larger, more extensive woods in the south of the district (Walker 1985).

Forest technology has only recently allowed the draining and ploughing of heavy, water-saturated peat (Thompson 1979, 1984). Thus it can be confidently assumed that very few long-established plantations were planted on peatlands. Such woodlands have a history of several hundred years and would have been established or developed on mineral soils. Figure 4.3 shows recent plantations to be concentrated in west Caithness and east Sutherland and on the peatlands around Loch Shin and Lairg.

Past and current extent of peatlands in Caithness and Sutherland

The 1:25,000 mapping allowed the calculation not only of the total area and distribution of flat, wet peatland suitable for moorland waders but also of the extent of peatlands before forestry resulted in major losses of blanket bog. Study of the underlying landforms shows nearly all recent forests to have been established on peat. Within the two districts there has been relatively little recent conversion of moorland to agricultural use, although this has been more significant in the distant past. Including newly afforested land, it is

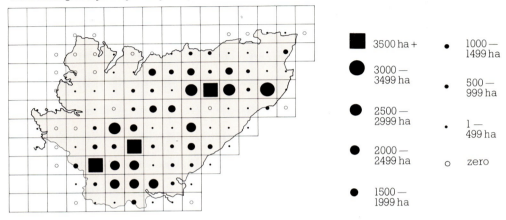

Figure 4.3 Extent of recent forestry plantations in Caithness and Sutherland in each 10km grid square (January 1986)

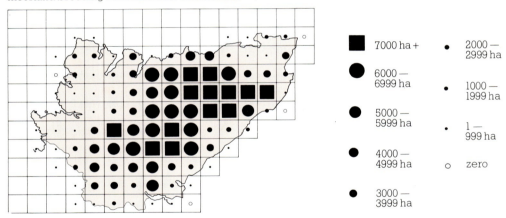

Figure 4.4 Estimated extent of original peatland cover suitable as habitat for moorland breeding waders before afforestation

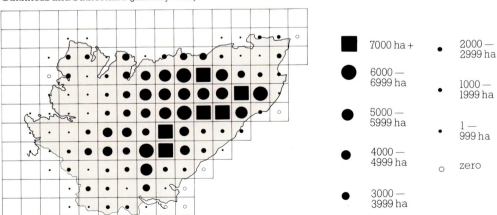

Figure 4.5 Extent of peatland suitable as breeding habitat for waders in Caithness and Sutherland (January 1986)

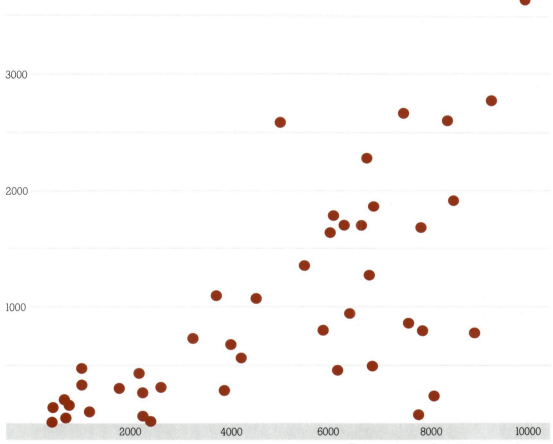

Figure 4.6 The relationship between the estimated area of original peatland cover and the area of recent forestry plantations.

Grid squares with substantial areas of sea and agricultural land are excluded

calculated that formerly the total area of peatland suitable as habitat for breeding waders (categories A, B and C) was about 343,000 ha. By August 1985 this area had fallen to some 270,100 ha. However, the loss of peatland to forestry has not been uniform over the whole region. There has been a much greater proportional loss of the lower-altitude (below 250 m) peatlands in north-east Sutherland and west Caithness (Angus 1987).

The total original area of blanket bog in Caithness and Sutherland has been calculated as about 401,375 ha (Lindsay et al. in prep.). This contrasts with the area of 343,000 ha calculated as originally suitable for waders, indicating that there is a considerable area of steeper mountainside with shallow or dissected peatland unsuitable for waders (included in category D).

Figure 4.5 shows the present

distribution of peatland suitable for breeding waders. Comparison of the distribution of forestry plantations (Figure 4.3) with that of the original peatland extent (Figure 4.4) shows that virtually all recent plantings have been on peatland suitable for waders. Areas unsuitable for moorland waders have been little planted. To the west of the Moine Thrust (Figure 1.3) there is a large area of land less favoured by moorland waders (see section 2.3). The same features (rockiness, steepness, fragmentation and, in places, high altitude) also make it largely unsuitable for the establishment of forests. To the east, the area of unsuitable land diminishes and, correspondingly, the extent of forestry plantations increases. There is a strongly significant positive correlation between the area of original peatland cover and the area of forest plantations in each 10 km grid square ($r = 0.691$, $n = 41$, $p<0.001$) (Figure 4.6). There is a corresponding significant negative correlation between the area of land unsuitable for waders and the amount of forestry in each 10 km grid square ($r = -0.693$, $n = 41$, $p<0.001$). This implies that both foresters and breeding waders select those areas which are flat and boggy, and so they are in direct competition for the same areas of land. This is the root of the current problem.

This coincidence of selection is due to waders favouring the wettest areas, which tend to be large and flat with well-developed pool systems. Foresters also prefer areas which are large and flat. This allows greater ploughing efficiency and less time is taken up in turning the ploughing rig at the end of each furrow. There are also other considerable economies of scale in afforesting uniform terrain, such as those associated with fencing. Pool systems are not usually initially drained and planted, but they become isolated when they are closely surrounded by forest, and their hydrology may be altered so that they can subsequently be afforested.

4.4 The use of landforms to estimate abundance and losses of waders on peatlands

Estimation of wader population sizes

The total area suitable for breeding waders was divided into three of the four landform categories outlined in section 4.2 and Table 4.3. Golden plovers, dunlins and greenshanks do not breed in every 10 km square in Caithness and Sutherland (Figure 3.4a-c) despite the presence of apparently suitable habitat, but they occur on most peatland areas. The mean densities found in each of the three landform categories takes such non-uniformity into account by including sites which appear suitable on habitat grounds but where species are either absent or breed at anomalously low densities. Inclusion of such sites will reflect the scarcity or non-occurrence of peatland waders in a few such areas of Caithness and Sutherland.

The area of each of the three landform categories was calculated within the area of land classed as suitable for breeding waders (Table 4.5 and Figure 4.7). Multiplying these areas by the average densities of breeding waders found for each category provides population estimates for the peatlands of Caithness and Sutherland (Table 4.5).

Golden plover

A total of 3,980 pairs of golden plovers is estimated to breed on the remaining unplanted peatlands in Caithness and Sutherland. The total British breeding population is estimated at about 22,600 pairs (data collated for Piersma 1986); thus the Caithness and Sutherland peatlands hold some 18% of British breeding golden plovers and 17% of the population breeding within the European Communities' territories (Table 8.1).

The breeding distribution in Figure 3.4a shows that golden plovers avoid agricultural land in the extreme north-east of Caithness and have a strong affinity for peatland. Unlike dunlin and greenshank, golden plover densities are often greater on slightly

Table 4.5 Estimated breeding populations of moorland waders on the Caithness and Sutherland peatlands based on associations of densities with landform categories
See Table 4.3 for definitions of the landform categories.

	Extent (km^2)	Density (pairs/km^2)	Estimated numbers (pairs)	Rounded total numbers (pairs)
Golden plover				
Landform A	828·5	2·37	1963	
Landform B	992·8	1·46	1449	
Landform C	880·1	0·64	563	
				3980
Dunlin				
Landform A	828·5	3·90	3231	
Landform B	992·8	0·57	566	
Landform C	880·1	0·04	35	
				3830
Greenshank				
Landform A	828·5	0·64	530	
Landform B	992·8	0·08	79	
Landform C	880·1	0·02	18	
				630

eroded peatlands where there is an even spacing of small hags and hillocks. The highest densities coincide with the area of greatest forestry expansion (Figure 4.3), thus giving a potential for further severe losses if afforestation of peatland habitat continues at its current rate.

Dunlin

A total of 3,830 pairs of dunlins is estimated to breed in Caithness and Sutherland. This estimate is higher than those made previously for the same area owing to a reanalysis of data using different methods (see the Appendix for details). This reanalysis gives a more realistic total for the two districts. Adjusting the data collated for Piersma (1986) by incorporating this figure gives a national total of 9,900 pairs. Thus Caithness and Sutherland hold about 39% of the British breeding population of dunlins and 35% of the European Communities' breeding population (Table 8.1). The methods used to survey for dunlins usually result in underestimates of true numbers, but this will apply generally so that relative proportions will be correct.

The highest breeding densities in Caithness and Sutherland show close agreement with the extensive wet areas of peatland (Figure 4.5). This species is thus one which would be directly and significantly affected by loss of further peatland habitat in Caithness and Sutherland.

Greenshank

A total of 630 pairs of greenshanks is estimated to breed in Caithness and Sutherland on landform categories A, B and C. The total British breeding population is currently estimated at 960, with most of the remainder being in Ross and only small numbers breeding elsewhere (Sharrock 1976). Thus Caithness and Sutherland hold about 66% of the British (and therefore the European Communities') breeding population of this species (Table 8.1).

The breeding distribution of the greenshank throughout Caithness and Sutherland is shown in Figure 3.4c. Although it is widespread in the extreme west of Sutherland (D. & M. Nethersole-Thompson 1979), the habitat here is fragmented and discontinuous because of interruption by unsuitable high mountains. In Caithness and east Sutherland the breeding habitat is much more

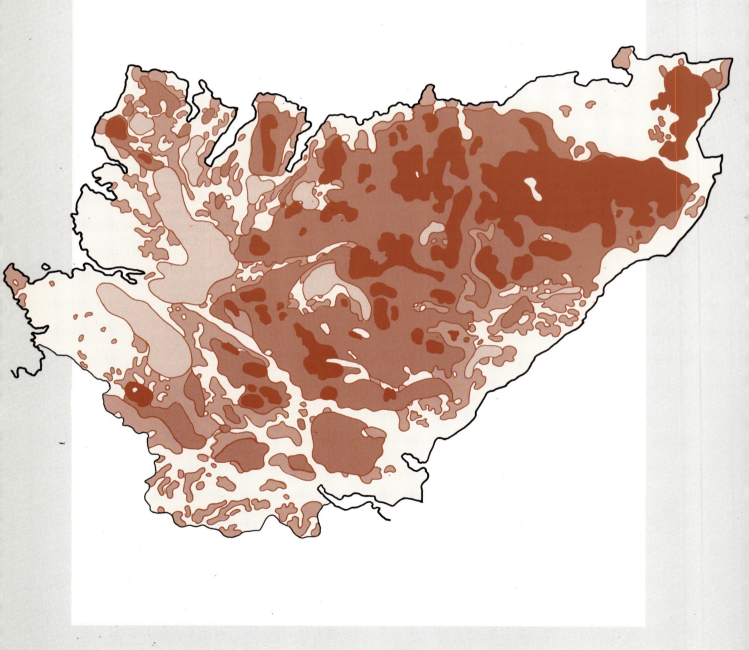

Figure 4.7 Extent and quality of habitat for breeding moorland waders on the Caithness and Sutherland peatlands, shown as the landform categories of Table 4.3

Key

Category A: Pool complexes and wet *Sphagnum* flows
Category B: Sloping blanket bog with pools
Category C: Steeper and more broken ground
Category D: Montane and other unsuitable areas

continuous. Greenshanks avoid the agriculturally modified land in the north and east of Caithness and are strongly associated with peatlands throughout both districts.

In a review of birds potentially affected by afforestation published by the Royal Forestry Society, Harris (1983) categorised the greenshank as breeding usually on intractable, wet boggy open spaces above the planting limit; this is totally incorrect and misleading. In Caithness and Sutherland, as elsewhere, a high proportion of the total British population is directly at threat from further loss of peatland habitat.

Losses of breeding waders

In the same way that habitat/density associations have been used to calculate the numbers of breeding waders currently present on the Caithness and Sutherland peatlands, the losses of breeding waders due to the afforestation of peatland are estimated in Chapter 6.

4.5 The numbers of other peatland breeding birds

The previous section has shown that it is possible to estimate numbers of golden plover, dunlin and greenshank based on their characteristic association with certain habitat and landform types. For the other peatland birds there are varying difficulties in estimating total population size, previous losses to afforestation or areas which are especially important. For some of the rare species and those favouring localised habitats which are shown on, and therefore easily located from, detailed maps (e.g. pool systems, larger lochs or crags), there are already counts of a large proportion of the total population. Special surveys of arctic skua, black-throated diver, peregrine and golden eagle have given good census information about total breeding numbers.

Other species are patchily distributed according to the occurrence of specialised habitats which are not readily identifiable from maps (e.g. snipe, ringed plover and redshank), and yet others are evidently widespread but elusive (e.g. merlin and short-eared owl). The rarest species, especially of waders, are extremely difficult to find, and in such a large area it is unlikely that all breeding pairs have been discovered. The 77 surveyed sites may give a reasonable sample from which the total numbers of some of these other species could be estimated within broad limits, but there is no means yet of testing such an assumption.

4.6 Summary of the ornithological interest

The outstanding features of the ornithological interest of the Caithness and Sutherland peatlands are the high species diversity and the large populations of breeding waders. No fewer than 15 species of waders are known to nest in the region and these include 66%, 39% and 18% of the total British breeding populations of greenshank, dunlin and golden plover respectively. There is a wider ecological spectrum of breeding birds, including waterfowl, raptors and passerines, than for any other moorland area in Britain. Important fractions of the total British breeding populations of other species are as follows: red-throated diver (14%), black-throated diver (20%), greylag goose (wild stock — 43%), wigeon (20%), common scoter (39%), hen harrier (5%), golden eagle (6%), merlin (5%), peregrine (5%), common gull (10%), and short-eared owl (5%).

Rare and local species are well represented, there being three species (Temminck's stint, ruff and wood sandpiper) each with 1-10 pairs nesting in Britain, two species (common scoter and red-necked phalarope) with 10-100 pairs nesting in Britain and seven species (black-throated diver, greylag goose, wigeon, hen harrier, golden eagle, merlin and peregrine) with 100-1,000 pairs nesting in Britain.

For 11 species, the area contains significant fractions of the total EC breeding populations, as follows: red-throated diver (14%), black-throated diver (20%), wigeon (20%), common scoter (16%), hen harrier (1%), merlin (4%), golden plover (17%), dunlin (35%), greenshank (66%), arctic skua (2%) and short-eared owl (4%).

Several species have declined and/or are still declining elsewhere in Britain — wigeon, buzzard, golden eagle, merlin, red grouse, golden plover, dunlin, snipe, curlew, greenshank, red-necked phalarope and raven. Some of these have already been reduced in numbers through afforestation in other districts as well as in Caithness and Sutherland.

Many of the above species are mainly or wholly northern European (boreal—arctic) in distribution and depend in the rest of their range on naturally treeless open wetlands and tundras. Britain supports the southernmost populations of these birds because of the large extent of open moorland resembling these more northern habitats. Caithness and Sutherland are an especially favourable area for this bird assemblage because the conjunction of climate and topography have given large areas of wet blanket bog, with a wide variety of associated open water habitats which simulate tundra. Some of the characteristic breeding birds of northern tundra are different, however, the goose tribe being represented in Caithness and Sutherland only by the greylag and the whimbrel being replaced by the curlew, so that the precise combination of species is not exactly replicated anywhere else in the world.

In the winter the region remains important for several scarce or local bird species. The peatlands are used as feeding habitat and roosting sites by internationally significant numbers of Greenland white-fronted geese. Golden eagles and hen harriers stay to hunt the moors, and the red grouse population is resident.

Unsurveyed habitats and species groups 5

5.1 Lochans, lochs and rivers

The study of open water habitats in Caithness and Sutherland for their botanical and invertebrate interest is only just beginning. Most of the standing open waters belong to the dystrophic or oligotrophic types, but little is known of their limnology at present. Caithness and Sutherland hold important freshwater fisheries: such rivers as the Helmsdale, Shin, Thurso and Naver have long been famous for the exceptional quality of their salmon. The streams and larger moorland lochs are also fished a good deal for brown trout. Whilst recent declines in catches on the Helmsdale and other rivers have been, in part, attributed to offshore netting, Egglishaw, Gardiner & Foster (1986), Graesser (1979) and others have demonstrated the close relationship between declines in salmon catches and the afforestation of upland catchments used as nursery streams.

5.2 Invertebrates

Because of the remoteness of the region the terrestrial invertebrate fauna is very little known and more survey work is needed to identify the range of species and record their abundance and distribution. Because of their geographical position, it is likely that the peatlands of Caithness and Sutherland support distinctive invertebrate communities differing in precise species composition from those found on peatlands further south in Britain.

Top left: Emperor moth
Bottom left: Northern eggar moth
Top right: Large heath butterfly (northern form)
Bottom right: Fox moth

They contain the most extensive population of the large heath butterfly *Coenonympha tullia* in Britain, essentially a peatland species, and large day-flying moths such as the emperor *Saturnia pavonia*, northern eggar *Lasiocampa callunae* and fox *Macrothylacia rubi* are numerous. The available evidence suggests that the lochs and pools within the peatlands support assemblages of freshwater invertebrates of conservation importance, including some species which are nationally rare. The arctic-alpine dragonfly *Aeshna caerulea* is known from the region and the recent discovery here of the water beetle *Oreodytes alpinus*, new to Britain and found elsewhere only in Siberia and the extreme north of Europe, suggests northern tundra affinities in the invertebrate as well as the ornithological fauna (Foster & Spirit 1986). Other nationally rare water beetles occur in the dubh lochans (Spirit 1986; Foster 1987).

Effects of afforestation on the ecosystem 6

The Nature Conservancy Council's report *Nature conservation and afforestation in Britain* (1986) gave a general account of the impact of afforestation on both the biological and the physical environment. Ratcliffe (1986) has also considered the effects on wildlife in upland Scotland. These commentaries are relevant to the Caithness and Sutherland peatlands, but the following statement makes more specific reference to the effects on this particular area, and especially on its birds. The impact of afforestation is complex, for not only are there obvious and direct effects on the ground ploughed and planted, but unplanted ground and freshwater habitats both within and beyond the forest are affected in varying degrees. In the situation of continually expanding afforestation, open ground steadily contracts in area and may become increasingly surrounded and 'squeezed' between blocks of forest, so that its relative value may change.

6.1 Effects on birds

The direct effect of afforestation is to replace the peatland bird assemblage with a woodland bird assemblage. Different species are affected at varying rates: some, such as dunlin, disappear quickly, while others may linger for several years, though usually in reduced numbers (Reed 1982a). The relevant outcome is that, by the time the young forest closes to the thicket stage at 10-15 years, the habitat has been transformed and is useless to most open moorland birds. They disappear and, at the minimum, their loss is directly proportional to the area and quality of ground planted. The problem is exacerbated because the flows are particularly attractive to forestry interests because of low land prices and the relative ease and cheapness of planting operations on such flat and unbroken ground. These flat flows are also the prime peatlands for both vegetation and breeding birds. There has thus been a selective conversion of many areas of the highest quality bird habitat to forest.

It is exceedingly unlikely that the birds displaced from planted ground are permanently absorbed by adjoining unplanted peatland, giving an increase in population density there. Any lasting absorption of displaced birds would be contrary to all that has been learned about the relationships between the carrying capacity of moorland and population densities of breeding birds. Results from one site in Sutherland surveyed before the afforestation of adjacent moorland showed that numbers of some waders increased after major planting but decreased in subsequent years to densities below the original levels. This indicates that, whilst birds were initially displaced from ploughed areas and attempted to breed close by, the population failed to maintain itself at this larger size. Although some species are subject to marked annual fluctuations, many are relatively constant, and all have an upper limit imposed by the productivity of the habitat. Most of the peatland bird species already have a non-breeding surplus which is excluded from nesting by the territorial behaviour and spacing of the breeding population.

Thompson, Thompson & Nethersole-Thompson (1986) found that, during years of high greenshank breeding density, egg weight was less and interactions between females more frequent (both factors militating against breeding performance) than when density was low. They concluded that "loss of habitat surrounding lochs and rivers used by feeding greenshanks will inevitably promote competition, increased predation, reduced food intake and, ultimately, the production of less viable eggs (through birds laying later and lighter clutches)". During their movement from nest sites to feeding areas, greenshank chicks (like other wader young) are at considerable risk from being trapped within the deep drainage ditches and plough-lines (Väisänen & Rauhala 1983). Nethersole-Thompson & Watson (1974) have documented the decline and extinction of groups of breeding greenshanks which required extensive open areas within the mature pine forests of Strathspey. When these were planted and dense cover grew up, the greenshanks deserted the area (see section 3.3).

There are, indeed, indications that afforestation leads to a thinning-out of breeding bird populations on adjoining unplanted moorland beyond the forest boundary or forming enclaves within it. For some species, this is only to be expected. There is typically a cessation of moor-burning on ground adjoining forest, and in some areas grazing is reduced as well through removal of sheep. This leads to an increase in luxuriance and age of vegetation, and after about ten years the effect is often quite marked, to the point where suitability as breeding habitat has declined for those species which need short vegetation, for example golden plover and dunlin. As an example, in Galloway scattered pairs of dunlins and a relatively dense population of golden plovers bred during 1950-60 on the moorlands between Airie and Grobdale, east of Loch Skerrow. By 1986, although some 20 km^2 of open moorland still remained, it was surrounded by large expanses of well-established forest: the dunlins had disappeared and the golden plovers had declined to low density (A. D. Watson pers. comm.).

Preliminary studies from moorland bird surveys in Caithness and Sutherland and also northern England suggest that some breeding species tend to avoid the forest edge, so that there is a zone of lower density on the moorland around the perimeter (Langslow 1983; Stroud & Reed 1986). Such an effect could be partly the result of habitat change, but it could also be a response to increased predation associated with the forest. In Kincardine, R. Parr (pers. comm.) found that afforestation of half of Kerloch Moor was followed not only by a substantial decline in the breeding population of golden plovers but also by a marked increase in nest predation for the residual breeders on the remaining moorland.

Afforestation creates new nesting habitat for carrion and hooded crows

Afforestation results in both the direct loss of patterned bogs and less obvious, long-term effects. The loss of such wetland areas has seriously affected internationally important moorland breeding wader populations. Loch More wetlands, Caithness

(Petty 1985) and much increased cover for foxes (Hewson & Leitch 1983). These predators are provided with secure refuges by the forest, for in thicket plantations their breeding places are usually impossible to locate; yet both rely largely on ground outside the forest for food and will perforce concentrate their search over the adjoining bogland. Until recently many parts of these peatlands, especially the remoter areas, were subject to only a low nest predation pressure from crows, for large areas were devoid of their nesting sites (in trees) and were distant from the nearest breeding pairs. Though crows range widely over the open moorlands, the advent of large blocks of forest in numerous places is bound to increase their penetration of the remaining open ground. The effects could be even more serious for the other waterfowl than for the waders and other species, since the vicinity of lochs and rivers is a focus for their nests and the young are taken there soon after hatching and so are readily located by predators.

These suggestive pointers have led to more detailed research, now in hand, on potential avoidance and predation responses which could represent a serious 'edge-effect' of afforestation, in addition to its primary impact in causing species loss. These effects may not appear until the plantations are well-grown, and they may become enhanced as the overall balance of areas and pattern of configuration between forest and peatland change through the advance of further new planting. In a study of birds of enclosed areas of moorland surrounded by conifer forests in southern Scotland, Rankin & Taylor (1985) concluded that the size of such areas and the number or diversity of habitats within the moorland 'islands' together had a dominant influence on the numbers and densities of breeding bird species. Only areas greater than 270 ha sustained a representative moorland bird assemblage including less common species such as merlin, ring ouzel and short-eared owl. The predatory birds need a much larger open moorland hunting area than that which will satisfy the feeding and other requirements of their prey species.

In areas of low carrying capacity, each pair of golden eagles may need to hunt over open moorland averaging 10,000 ha in area. In Galloway, the reduced breeding performance of three out of four pairs of golden eagles was correlated with an average 43% afforestation of their feeding areas by 1979 (and an average 69% of the more productive ground below 905 m). One pair, which subsequently disappeared, had lost 62% of their feeding area and 90% of their low ground (Marquiss, Ratcliffe & Roxburgh 1985). Afforestation also accounted for most of the 56% decline in the raven breeding population between 1946-60 and 1974-76 in southern Scotland and northern England reported by Marquiss, Newton & Ratcliffe (1978). Raven decline here has continued in parallel with further planting, reaching 72% by 1981 (Mearns 1983). In some parts of these regions the decrease has been proportional to the area afforested, but some raven territories have been deserted even when they are only partly planted. In Galloway again, a hill population of buzzards numbering at least 25 pairs in 1946-55 had declined to only two pairs in 1981 (D. A. Ratcliffe unpublished; Mearns 1983): although only about 50% of the area had been afforested, a much higher proportion of the more productive low ground had been lost.

The greater ability of ravens and buzzards in mid-Wales to maintain their numbers in the face of afforestation appears to depend on the exceptionally good food supply of the remaining open ground in this area and the less continuous distribution of blocks of forest (Newton, Davis & Davis 1982). Forest edges have provided tree nest sites for a few additional pairs of ravens and buzzards in parts of Wales where both tree and rock sites were otherwise lacking. However, for these species and other moorland raptors such as merlin and kestrel which may utilise old crows' nests within the new forests, the crucial issue is the extent of open hunting

6

ground remaining beyond the plantations, since the new forests have little or no value as feeding habitat.

Open ground remaining within forests as rides and roadside verges is virtually useless as nesting habitat to nearly all the birds of the Caithness and Sutherland peatlands. The nests of larger species would be far too readily located by predators (Ratcliffe 1986). Apart from occasional pairs of small passerines, these areas are devoid of nesting birds. Common sandpipers may continue to nest on stream banks where the forest edge is kept well back, but most waders need a broader expanse of nesting habitat. Even where patches of pool and hummock 'flow' are left unplanted within the forests, they can be of only doubtful value to breeding waders and other waterfowl. Mostly they are too few and too small, and habitat changes resulting from the abruptly abutting forests are probable (Chapman & Rose 1986).

6.2 Losses of peatland birds

The methods used to estimate the quality of breeding habitat and hence the overall numbers of breeding waders on existing moorland (see Chapter 4) can also be used to estimate the losses of these on areas recently afforested. The areas of recent planting and land released for planting were superimposed on earlier, pre-afforestation Ordnance Survey maps, and the quality of lost peatland habitats assessed according to the four landform categories of Table 4.3. It was then calculated that 912 pairs of golden plovers, 791 pairs of dunlins and 130 pairs of greenshanks once used moorland occupied or planned to be occupied by plantations (Figure 1.4). On this basis, the original, pre-afforestation populations for the Caithness and Sutherland peatlands of Figure 1.2 can be calculated to have been 4,900 pairs of golden plovers, 4,620 pairs of dunlins and 760 pairs of greenshanks. There has thus been an actual or predictable loss of 19% of golden plovers, 17% of dunlins and 17% of greenshanks as a direct effect of afforestation. It has not been possible yet to estimate the losses of other breeding bird species, but most of those listed in Table 6.1 can be

Table 6.1 Declines in bird species due to afforestation in other parts of Britain (Sources: Marquiss *et al.* 1978, 1985; Mearns 1983; Nature Conservancy Council 1986; D. A. Ratcliffe pers. comm.)

Species	Area
Wigeon	Ettrick District
Greenshank	Spey Valley
Dunlin	Wales, Cheviots, Southern Uplands of Scotland (at least 400 pairs lost)
Golden plover	Wales, North York Moors, Cheviots, Southern Uplands, Eastern Highlands (at least 2,000 pairs lost)
Curlew	Wales, Northern England, Southern Uplands, Highlands (several thousand pairs lost)
Snipe	Wales, Northern England, Southern Uplands, Highlands (no estimate)
Red grouse	Wales, Northern England, Southern Uplands, Highlands (no estimate, but must be thousands of pairs lost)
Golden eagle	Southern Uplands, West Highlands (15-20 pairs lost)
Merlin	Wales, Northern England, Southern Uplands, Highlands (tens of pairs lost)
Buzzard	South-west Scotland (*c.* 25 pairs lost)
Raven	Cheviots, Southern Uplands, South-west Highlands (tens of pairs lost)

In addition, species such as lapwing, redshank, curlew, snipe, whinchat and stonechat have declined in the lowlands through agricultural development, so that their upland populations are now increasingly important.

presumed to have been affected.

Some of the Caithness and Sutherland peatland birds which occur widely in the British uplands have already lost a good deal of ground elsewhere, through the widespread afforestation of their habitats on both blanket bog and drier moorland (Table 6.1). The national populations of golden plover, dunlin, curlew, red grouse and merlin have all been significantly reduced by this transformation of habitat. These losses elsewhere are almost certain to continue through still further afforestation, so that the Caithness and Sutherland populations of affected species will become an increasing proportion of the British totals — unless afforestation continues here, too, at the present rate.

The pre-thicket stage after afforestation is beneficial to some birds which nest in fairly dense cover, including some open moorland species such as short-eared owl, hen harrier and black grouse. This phase is ephemeral and 'once-off', however, and cannot influence the long-term assessment of impacts. The crucial question is whether the cycle of return to open ground, when the forest is harvested and before the next generation of trees again closes to thicket, can restore a sufficiently large and close approximation to the original peatland bird community. Given that the forest rotation will, in the longer term, maintain a certain proportion of pre-thicket ground at any one time, foresters have argued that it will continue to maintain a parallel representation of moorland birds (e.g. Garfitt 1983).

As a general thesis, this argument has been found wanting (Nature Conservancy Council 1986; Ratcliffe 1986): the original bird community is not restored, and there is a shift to a mixed type in which some of the original species return, but less numerously than those of open woodland, scrub or forest edge and glade. While it will be some time before the situation on the Caithness and Sutherland peatlands can be properly observed, the effect of afforestation in drying out the ground makes it even less likely that there will be any significant return of the waders regarded as so important in this area. Observation of open areas created by large-scale death of young trees from pine beauty moth attacks does not suggest that anything approaching the peatland bird community redevelops.

6.3 The lack of compensatory gain in forest birds

Nature conservation and afforestation in Britain (Nature Conservancy Council 1986) examined in depth and rejected the arguments that afforestation replaces moorland by a woodland habitat with a richer bird fauna and that it restores a more original type of habitat to a man-made landscape. A main conclusion of this report is as follows (p. 64):

"When the heaths, grasslands, peatlands or sand dunes that are lost to afforestation have special nature conservation value in their previous state, the forests that replace them, however good they may become for wildlife, are not, and never can be, an adequate substitute. This is so because they can never contain more than a fragmentary, depleted and therefore inadequate representation of the open ground habitats, communities, species and physical features in their former wholeness."

Afforestation has a particularly destructive effect on wet ground habitats. The forests that are created are quite different from natural boreal coniferous forest; not only are they almost wholly composed of exotic species, but their structure and lack of subsidiary vegetation layers are highly artificial and will remain so. Whatever opportunities may exist for diversifying conifer plantations on fertile, lowland sites in more southern districts, they are minimal in Caithness and Sutherland, where most plantations will remain unthinned until they are felled because of the very high risks of wind-throw. The still more important point, however, is that the blanket bog covering so much of the region is a naturally treeless climax vegetation developed under the extremely oceanic climate and that

6

Mature conifer forests are dense and dark. These even-aged plantations lack the structural diversity found in natural woodlands. The forest birds encouraged by such plantations do not compensate for the loss of the moorland birds displaced

forest of any kind on the flows is an artefact (see section 2.2).

It has been claimed by forestry interests that afforestation in Caithness and Sutherland is beneficial to the conservation of birds because:
- it increases the numbers of individuals, or even of species, compared with those living on the peatland habitats;
- it enhances overall diversity, giving the best of both worlds, since some unplanted peatlands will still remain.

These arguments are invalid. The bulk of the resulting forest bird fauna consists of small passerine songbirds which are common throughout Britain, many of them in hedgerows and even in suburban gardens (Table 6.2). Their occurrence in greater numbers than the displaced birds of the peatlands is irrelevant to the conservation issue. By the same token, the additional "diversity" created through limited afforestation is of little account if the bird community which is added has low intrinsic conservation value. The amount of forest which has already been created is quite sufficient to give *de facto* representation to this viewpoint: with any more, the balance-sheet of losses compared to gains becomes increasingly adverse and unacceptable. Even if rarer songbirds such as crossbill, redwing, fieldfare and brambling colonise the forest, there is more than enough of such habitat already and this does not provide an argument for further forest expansion. It has to be remembered that, for some peatland and moorland birds, this is not the only region where contraction and loss are occurring through afforestation.

Table 6.2 Bird species found in coniferous plantations in Caithness and Sutherland

Species	National population estimates (pairs)	Species	National population estimates (pairs)
Buzzard	8,000—10,000	Goldcrest	500,000
Kestrel	30,000—40,000	Spotted flycatcher	300,000
Woodcock	8,000—35,000	Long—tailed tit	200,000
Tawny owl	50,000—100,000	Coal tit	500,000
Wren	3,000,000—5,000,000	Blue tit	3,500,000
Dunnock	2,000,000	Great tit	2,000,000
Robin	3,500,000	Treecreeper	200,000—300,000
Redstart	140,000	Carrion/hooded crow	1,000,000
Whinchat	15,000—30,000	Starling	3,000,000—6,000,000
Blackbird	4,500,000—5,000,000	Chaffinch	5,000,000
Song thrush	1,500,000	Greenfinch	800,000
Mistle thrush	300,000	Siskin	15,000—30,000
Sedge warbler	200,000	Scottish crossbill	500
Willow warbler	2,500,000	Reed bunting	400,000

6.4 Effects on vegetation

Afforestation causes the replacement of peatland vegetation by dense stands of conifers with few other plants. During the pre-thicket stages some of the existing species, especially grasses, some sedges and dwarf shrubs, become more luxuriant and tussocky. The deep ploughing and draining used as standard ground preparation rapidly dry out the surface of the intervening peat ridges and cause a loss of the strongly moisture-loving species. When the forest closes to thicket, virtually all the ground vegetation disappears, because the shade is so intense and the litter fall so heavy. In these unthinned forests this condition will persist until clear-fell harvesting, and this is a main reason why they can never be regarded as equivalent to natural boreal forests. The vegetational interest of these new plantations is thus limited to whatever ground is left unplanted and whatever cover redevelops during pre-thicket phases of subsequent rotations.

Ratcliffe (1986) has given reasons for regarding the vegetation of linear open habitats such as rides, roads and streamsides as a particularly fragmentary, uncharacteristic and unsatisfactory representation of the range of open moorland communities. This is even more the case for peatlands because of the pronounced drying effects which result. The only residual open habitats worth considering in this context are wider areas, mainly patches of pool and hummock flow, left unplanted within the forests. The questions to be asked about these are whether they are a sufficient representation of the peatland ecosystem and whether they will retain their previous character in the longer term.

Many of the true boreal forests elsewhere naturally contain open areas of bog too wet for trees to grow on. They range in size from tiny pockets to huge expanses, but most of them result from underlying topography and soil which

Figure 6.1 Diagrammatic representation of natural Scandinavian forest bog and afforestation blanket bog in Scotland to illustrate the significant differences.

Whilst underlying topography may be similar, blanket bog swathes the mineral substrate, contrasting with isolated basins of peat in Scandinavia. Forest bog trees are of mixed age and size, and waterlogging on the bogs causes a gradual transition, with increasingly stunted trees around the edges of the basins. The open structure of these forests reflects their ecological maturity. Plantations in Caithness and Sutherland consist of dense, uniform trees of even age which fragment once continuous blanket bog and cannot be thinned to achieve a more 'natural' state owing to the severe risks of wind-throw

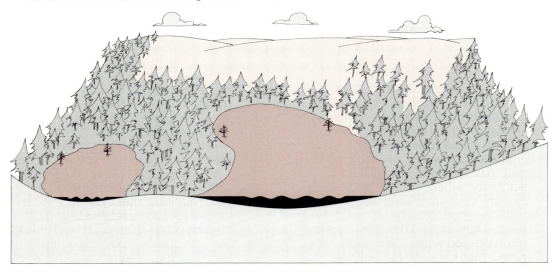

Scandinavia: dry climate, with raised bogs developed only where topography maintains a high water-table and with forest on dry mineral soils and bog margins

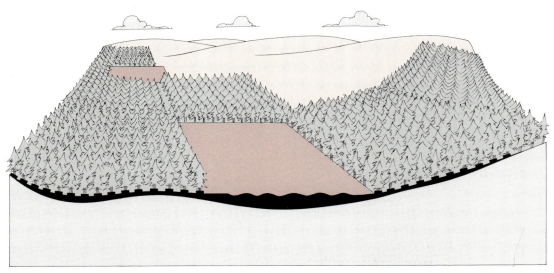

Scotland: wet climate, with blanket bog developed everywhere except on steep ground and with plantations established on peat where ploughing and draining have lowered the water-table

give impeded drainage. Most of these forest bogs are accordingly classified either as valley mires and fens or as raised bogs. Others are types not found in Britain such as aapa mires and palsa mires. Where the ground slopes steeply into wetter places, there may be a clear-cut edge between forest and bog, but more typically there is a gradual transition in which the trees thin out and become stunted, appearing finally as an irregular perimeter of dwarf 'checked' growth on the bog edges where the water-table is critically high. Some bogs have a scattering of these checked trees thinly distributed over much of the surface, but others have a central treeless area. Natural forest bogs show enormous variety in their relationship to the surrounding forests (see, e.g., Moen 1985; Eurola & Holappa 1985).

Enclaves of wet flow within plantations may in time develop a superficial similarity to these natural forest bogs. In many cases there will be a hard, wall-like edge to the trees, but natural regeneration may in time produce a more graded edge. The main point to note is that, because the Caithness and Sutherland examples are blanket bog, the leaving of arbitrary unplanted 'islands' of flow will not approximate to any natural conditions, for their relationships with the surrounding body of peat are lost (Figure 6.1). While the pool and hummock systems may be of outstanding interest, their development has depended on the growth of peat over a wider surrounding area. This situation also makes residual 'islands' of flow vulnerable to further changes likely to reduce or destroy their remaining interest.

Even limited drainage can have very serious effects on the ecology of blanket and raised bogs (Ivanov 1981; Lindsay 1987). Not only can limited surface drainage such as moor-gripping affect the level of the water-table and the composition of the surface vegetation (Lindsay in prep.), but peripheral deep drainage can alter the ground-water 'mound' (Ingram 1982), resulting in a lowering of the bog surface at considerable distances from the ditch. The Irthinghead Mires, a Site of Special Scientific Interest in Northumberland, are a series of small but important bogs protected from direct afforestation but now totally surrounded by the Kielder Forest. Charman (1986) has correlated the decreased plant diversity on these remnant bog sites with increasing age of surrounding forestry and has speculated that this is a result of a falling water-table, though fertiliser or spray drift may also be involved.

One of these bogs, Coom Rigg Moss, became totally surrounded by conifer plantations in the 1950s and 1960s. Since that time there have been marked and major changes to the flora and structure of this peat bog, and, whilst the precise mechanisms are not yet clear, "it is extremely unlikely that the changes seen at Coom Rigg would have taken place if the surrounding area had not been afforested" (Chapman & Rose 1986). If the full conservation interest of ombrotrophic mire systems is to be preserved, complete hydrological systems need to be protected intact. In the northern flow country, such hydrological systems can be extensive, because of the breadth of watersheds and the extent of catchments.

The implications are of a high probability that isolated portions of flow left within planted forest will lose their ecological and botanical interest through inevitable consequent change.

The remaining possibility, that peatland vegetation will regenerate to its previous state during successive clearance phases in the future forest rotations, seems very unlikely. The deep-ploughing and draining required for afforestation cause a permanent drying of the affected bog surfaces (Pyatt 1987). Water-table and surface flow patterns are immediately disrupted (McDonald 1973; Robinson 1985) and these changes are followed by longer-term shrinkage of peat (Prus-Chacinski 1962), wastage and oxidation (Silvola 1986). As the trees become established, higher evapo-transpiration rates lower the water-table further, contributing to a complete change in soil structure and in ground flora. So when the forest is eventually felled, the habitat will never revert to its previous state, for the physical conditions are permanently altered. Even if there is regeneration from buried seed or recolonisation from outside, the ground becomes too dry ever to support again the moisture-loving plants which characterised the previous blanket bogs. Species associated with dry peat surfaces are likely to become dominant, and the probability is that a mixed grass-heath will develop widely. It is noticeable that in areas opened up through pine beauty moth 'kills', a great abundance of purple moor-grass *Molinia caerulea* and rosebay willowherb *Chamerion angustifolium* has rapidly developed, evidently in response to the high nutrient levels from added fertiliser.

Patterns of afforestation have caused the isolation and fragmentation of previously extensive bog systems. The vegetation of such isolated areas will slowly change owing to altered management practices and the 'edge-effects' of the new plantations. Forsinard, Sutherland

Finally, the deep-ploughing and progressive modification of the upper peat after afforestation (including erosion and redistribution of material) renders the affected areas worthless for studies of vegetational, climatic and land-use history through pollen analysis. The sub-fossil pollen and other plant remains which allow such historical reconstructions and correlations (Godwin 1975) have to exist in an undisturbed state, free from any suspicion of modification to the complete peat profiles within which they are buried and preserved. Surface oxidation destroys the upper layers of the record, while any redistribution of material confuses it hopelessly. Cracking of peat also allows material from the surface to sink deeply below. While it may be that a sufficient number of undisturbed profiles will survive in unplanted sites within the region, the precise needs for the location of sample sites within these palaeo-ecological studies cannot be predicted and, quite often, fairly closely spaced samples along transects are needed.

6.5 Effects on abiotic features

Afforestation also has several quite marked effects on the physical/chemical environment, and these have been recently reviewed (Nature Conservancy Council 1986). There has so far been insufficient research to establish their relative importance in Caithness and Sutherland, but some general extrapolations and predictions can be made. Ploughing on these peatlands has the immediate and inevitable consequence of increasing the amplitude of peak stream flows by shortening the duration of peak run-off; after rain the excess water runs off much more rapidly than before, and total water yield increases slightly (Robinson 1980). When the forest has closed to form thicket, there is a decrease in water yield through interception effects (Calder & Newson 1979). Erosion of peat and mineral soil is another inevitable result of ploughing (Robinson & Blyth 1982), and subsequent cracking of deep peat is reported (Pyatt & Craven 1979; Pyatt 1987). Run-off and stream sedimentation can be reduced by modifying the system of draining, but on deep peat this is difficult because the whole purpose of draining is to lower water-tables by removing more water than before. Recent research on the blanket bog plateau of Llanbrynmair Moors in Wales has shown that the implementation of management guidelines on water protection has failed to prevent an increase in stream sediment loads to 246% of former values in one catchment and to 479% of former values in another, after forestry ploughing (Francis 1987).

Drying of acidic peat increases its acidity by oxidation (Pearsall 1938; Silvola 1986). The acidification of stream waters by the 'scavenging' of atmospheric acidity by surrounding forest is now well known (e.g. Harriman & Morrison 1982). While the prevailing acidic peats, soils and waters of Caithness and Sutherland are vulnerable to enhanced acidification, the lower levels of atmospheric acidity in this region (compared with, for example, Galloway) are likely to moderate the degree to which such acidification occurs (Fry & Cooke 1984). Data should nevertheless be collected and the situation monitored. The other major chemical effect, of nutrient enrichment of peats, soils and waters by the addition of fertilisers to promote forest growth, is potentially serious in its biological implications. This can cause local eutrophication, leading to algal blooms in lakes and reservoirs, and the same effect has been observed after significant clear-felling of part of a catchment (Parr 1984).

Studies of these changes and their biological effects are patchy and, in some cases, fragmentary, and there is a particular dearth of information for Caithness and Sutherland. There is an accumulating literature to show that, in one area or another, the effects of afforestation on freshwater fisheries are generally adverse. This subject was reviewed in *Nature conservation and afforestation in Britain* (based on Graesser 1979; Harriman & Morrison 1982; Stoner, Gee & Wade 1984; Stoner

& Gee 1985; Drakeford 1979, 1982). Egglishaw, Gardiner & Foster (1986) have since shown a high degree of correlation throughout Scotland between extent of afforestation and decline in salmon catches. Acidification of lakes in Sweden has affected water birds (Eriksson, Henrikson & Oscarson 1980; Eriksson 1984), and a decline in dippers has been correlated with enhanced river acidification in Wales associated with afforestation (Ormerod, Tyler & Lewis 1985). These findings point to the need for study of the chemistry of the Caithness and Sutherland peatlands. An important aspect of effects on abiotic features is that they are mostly not confined to the ground actually planted, but have significant overspill effects into adjacent areas, especially in the case of rivers and lakes. Afforestation therefore has to be assessed in its effect on the whole catchment.

6.6 An overall assessment

Afforestation so fundamentally transforms the peatland ecosystem that it is effectively destroyed. The fauna and flora are lost and the physical attributes so altered that, even if forestry ceased, any reconstitution of the previous ecosystem would be for all practical purposes impossible. The pre-thicket stage of subsequent forest rotations can only be regarded as a derisory restoration of the peatland ecosystem. In view of the importance of the peatland ecosystem, its replacement by a type of forest which is artificial and present over increasingly large areas of Britain, with little regional variation, amounts to a substantial net loss of nature conservation interest. The various possibilities of effects extending beyond the forest perimeter can only compound this loss.

The rivers draining from the peatlands of Caithness and Sutherland support major salmon fisheries which are important to the local economy. Blanket afforestation has seriously affected such fisheries elsewhere

International implications 7

The international importance of the peatlands of Caithness and Sutherland has to be considered in relation to overseas opinion and the requirements which stem from international treaties concerning nature conservation to which the United Kingdom is a party.

The Northern Highlands of Scotland, which include this area, were identified as globally important and listed as a "priority biogeographical province of land for the establishment of protected areas" in the World Conservation Strategy (IUCN/UNEP/WWF 1980), a document welcomed by the UK Government. This lays great stress on the conservation of non-renewable resources, the preservation of genetic diversity and the maintenance of essential ecological processes and ecosystems. All these considerations are pertinent to the continued survival of peatlands in Caithness and Sutherland. Indeed, the UK non-governmental response to the World Conservation Strategy (World Wildlife Fund-UK et al. 1983) suggests that the results of upland bird surveys show that "forestry operations should be planned with great care".

In September 1986 the International Mire Conservation Group (IMCG) visited the peatlands of Caithness and Sutherland on a study tour. This group of peatland ecologists from nine countries considered the blanket bogs of the north of Scotland to be "unique and of global importance", expressed its "dismay at the extent to which afforestation was found to be destroying this internationally important habitat" and viewed "the speed of destruction with particular alarm" (International Mire Conservation Group 1986).

The United Kingdom is a signatory to four international treaties which are relevant to the conservation of peatland habitats — the 'Bern' Convention, the 'Ramsar' Convention, the EEC Directive on the Conservation of Wild Birds and the World Heritage Convention.

7.1 The 'Bern' Convention on the Conservation of European Wildlife and Natural Habitats

Article 3 of this Convention requires the promotion of "national policies for the conservation of wild flora, wild fauna and natural habitats, with particular attention to endangered and vulnerable species... and endangered habitats". Article 4 specifically concerns the conservation of habitats. Article 4(1) states that "appropriate and necessary measures [shall be taken] to ensure the conservation of the habitats of the wild flora and fauna species, especially those listed in the Appendices I and II, and the conservation of endangered natural habitats". Article 4(2) states: "The Contracting Parties in their planning and development policies shall have regard to the conservation requirements of the areas protected under the preceding paragraph, so as to avoid or minimise as far as possible any deterioration of such areas." Article 4(3) requires the special protection of areas important to breeding migratory species.

These provisions are applicable to the peatlands of Caithness and Sutherland. Their status as an endangered and important natural habitat has been recognised by the International Union for Conservation of Nature and Natural Resources (IUCN/UNEP/WWF 1980), Royal Society for the Protection of Birds (1985), IMCG (1986), NCC (1986 and this report), Ramblers' Association (Tompkins 1986), Scottish Wildlife Trust (1987), Council for the Protection of Rural England (Stewart 1987) and others. Many of the species listed in Appendices I and II of the Convention occur on these peatlands. These include otter, black-throated diver, red-throated diver, little grebe, merlin, peregrine, ringed plover, wood sandpiper, common sandpiper, Temminck's stint, dunlin, red-necked phalarope, short-eared owl, pied wagtail, grey wagtail, wren, whinchat and stonechat.

7.2 The 'Ramsar' Convention on Wetlands of International Importance especially as Waterfowl Habitat

The United Kingdom is also a party to the 1971 'Ramsar' Convention, having signed it in 1973 and ratified it in 1976. The preamble to this Convention, which specifically includes peatlands, indicates that the Contracting Parties are "convinced that wetlands constitute a resource of great economic, cultural, scientific and recreational value, the loss of which would be irreparable" and that they desire "to stem the progressive encroachment on and loss of wetlands now and in the future". Article 3(1) of the Convention requires that "the contracting Parties shall formulate and implement their planning so as to promote the conservation of the wetlands included in the list [of protected sites], and as far as possible the wise use of wetlands in their territory".

The Conference of the Contracting Parties held in Cagliari, Italy, in 1980 agreed eight criteria for the assessment of international importance of wetlands. A wetland should be considered internationally important if it met any of these criteria. The relevant text says:

"1. Quantitative criteria for identifying wetlands of importance to waterfowl. A wetland should be considered internationally important if it:
a. regularly supports either 10,000 ducks, geese and swans; or 10,000 coots; or 20,000 waders
or
b. regularly supports 1% of the individuals in a population of one species or subspecies of waterfowl
or
c. regularly supports 1% of the breeding pairs in a population of one species or subspecies of waterfowl.
2. General criteria for identifying wetlands of importance to plants or animals. A wetland should be considered internationally important if it:
a. supports an appreciable number of a rare, vulnerable or endangered species or subspecies of plant or animal
or
b. is of special value for maintaining the genetic and ecological diversity of a region because of the quality and peculiarities of its flora and fauna
or
c. is of special value as the habitat of plants or animals at a critical stage of their biological cycles
or
d. is of special value for its endemic plant or animal species or communities.
3. Criteria for assessing the value of representative or unique wetlands. A wetland should be considered internationally important if it is a particularly good example of a specific type of wetland characteristic of its region."

The Caithness and Sutherland peatlands are quite exceptional in meeting all eight of these criteria.

Criterion 1a: The peatlands of Caithness and Sutherland regularly support more than 20,000 waders.

Criterion 1b: The peatlands support at least 1% of the individuals in biogeographical populations of the following waterfowl: Greenland white-fronted goose, native greylag goose, dunlin and golden plover.

Criterion 1c: The peatlands support at least 1% of the breeding pairs of the following biogeographical populations of waterfowl: native greylag goose, dunlin and golden plover.

Criterion 2a: The peatlands support an appreciable number of individuals of several rare, vulnerable or endangered species or subspecies of plants and animals. Among the species or subspecies in this category are: black-throated diver, Greenland white-fronted goose, greenshank, wood sandpiper, merlin and golden eagle.

Criterion 2b: The ecological peculiarities of the peatland fauna and flora have been described in this report. The peatlands clearly qualify under this criterion.

Criterion 2c: As breeding and wintering habitat for large numbers of specialised peatland birds, insects and plants, the peatlands fulfil this criterion.

Criterion 2d: The peatlands qualify under this criterion since some of their plant communities and their breeding bird assemblage are globally unique.

Criterion 3: The global importance of the area as an outstanding example of oceanic blanket bog has been recognised by such groups as the International Mire Conservation Group (see above). This area is a supreme example of blanket bog development.

Contracting Parties to the 'Ramsar' Convention provide regular reports on wetland conservation. The UK report to the Conference of the Contracting Parties at Groningen, Holland, in 1984 stated: "Recent research has suggested that mire systems are the most threatened wetland habitat in Britain" and "commercial afforestation is now a major threat to blanket mire systems, promoted by tax incentives rather than the direct economic value of the timber produced" (IUCN 1984, p. 452).

7.3 EEC Directive on the Conservation of Wild Birds

As a member of the EEC, the UK has accepted and is bound by Directive 79/409 of 2 April 1979. The Directive is concerned with many aspects of bird conservation, but Article 4 concerning habitat protection is of particular relevance to the peatlands of Caithness and Sutherland. This says:

"1. The species mentioned in Annex 1 shall be the subject of special conservation measures concerning their habitat in order to ensure their survival and reproduction in their area of distribution.

In this connection, account shall be taken of:
a. species in danger of extinction;
b. species vulnerable to specific changes in their habitat;
c. species considered rare because of small populations or restricted local distribution;
d. other species requiring particular attention for reasons of the specific nature of their habitat.

Trends and variations in population levels shall be taken into account as a background for evaluations.

Member States shall classify in particular the most suitable territories in number and size as special protection areas for the conservation of these species, taking into account their protection requirements in the geographical sea and land area where this Directive applies.

2. Member States shall take similar measures for regularly occurring migratory species not listed in Annex 1, bearing in mind their need for protection in the geographical sea and land area where this Directive applies, as regards their breeding, moulting and wintering areas and staging posts along their migration routes. To this end, Member States shall pay particular attention to the protection of wetlands and particularly to wetlands of international importance.

3. Member States shall send the Commission all relevant information so that it may take appropriate initiatives with a view to the co-ordination necessary to ensure that the areas provided for in paragraphs 1 and 2 above form a coherent whole which meets the protection requirements of these species in the geographical sea and land area where this Directive applies.

4. In respect of the protection areas referred to in paragraphs 1 and 2 above, Member States shall take appropriate steps to avoid pollution or deterioration of habitats or any disturbances affecting the birds, in so far as these would be significant having regard to the objectives of this Article. Outside these protection areas, Member States shall also strive to avoid pollution or deterioration of habitats."

The following species listed under Annex 1, and thus requiring habitat protection under Article 4(1), occur on the peatlands of Caithness and Sutherland as either breeding or wintering species or subspecies: black-throated diver, red-throated diver, Greenland white-fronted goose, hen harrier, golden eagle, peregrine, merlin, golden plover, wood sandpiper, red-necked phalarope and short-eared owl. Most of the remainder of the breeding bird species, especially the waders, are migratory and require habitat protection under Article 4(2).

Under Article 4(4) Member States are required to strive to avoid pollution or deterioration of habitats both inside and outside protected zones. This is particularly relevant to the afforestation of peatlands since such afforestation results both in direct habitat loss (Chapter 6) and in pollution of watercourses and other wider effects (section 6.5).

7.4 The World Heritage Convention

The Convention concerning the Protection of the World Cultural and Natural Heritage, or the World Heritage Convention, which the UK Government ratified in 1984, requires each State Party to ensure that "effective and active measures are taken for the protection, conservation and preservation of the cultural and natural heritage situated on its territory" which qualifies for World Heritage status under the Convention.

The following are the criteria for "natural heritage":

"natural features consisting of physical and biological formations or groups of such formations, which are of outstanding universal value from the aesthetic or scientific point of view;

geological and physiographical formations and precisely delineated areas which constitute the habitat of threatened species of animals and plants of outstanding universal value from the point of view of science or conservation;

natural sites or precisely delineated natural areas of outstanding universal value from the point of view of science, conservation or natural beauty."

The peatlands of Caithness and Sutherland meet all three criteria, and IUCN, which is responsible for vetting "natural site" World Heritage submissions for Unesco and the World Heritage Bureau and making recommendations for the World Heritage List, has recently urged the Nature Conservancy Council to consider nominating the peatlands of Caithness and Sutherland as the first British wetland listed under the World Heritage Convention. This Convention recognises the concept of a "common heritage" and that certain unique "cultural and natural properties" constitute "a world heritage for whose protection it is the duty of the international community as a whole to co-operate".

Birds and bogs: their conservation needs   8

The Caithness and Sutherland peatlands should be viewed as an ecosystem in which the different elements — of physical environment, plants and animals — are a functionally interdependent whole. Whilst these elements are evaluated separately and this report places special emphasis on the birds, overall nature conservation value resides in this unity. The most significant aspects of nature conservation importance are as follows.

National value

- The large area and diversity of blanket bog as a physiographic/vegetation feature and the relative lack of disturbance in many places, giving the greatest extent of actively growing mire in Britain and one of the few areas of extensive natural terrestrial vegetation now remaining.
- The extensive development of patterned flow as a feature rare in British bogs elsewhere and the great variety shown by these pool and hummock systems.
- The abundance of certain rare or local bog plant species.
- The greater diversity of the breeding bird assemblage than that of moorland and bog elsewhere in Britain.
- The large total numbers of many bird species, considered as percentages of their total British populations (Table 8.1) and bearing in mind that some of these species are declining elsewhere and will continue to do so, especially as a result of afforestation.
- The presence of several nationally rare breeding bird species.
- The predicted equivalence of interest relating to habitats (mainly open waters) and groups (especially invertebrates) not yet surveyed.

International value

- One of the largest and most intact known areas of blanket bog (a globally rare ecosystem-type) in the world.
- A northern tundra-type ecosystem in a southern geographical and climatic location, by reason of the extreme oceanicity of the northern Scottish climate.
- Development of unusually diverse systems of patterned surfaces on blanket bog instead of on other types of bog/mire where their analogues occur elsewhere in the world.
- A floristic composition of blanket bog and associated wet heath vegetation unique in the world and representing a highly Atlantic influence on plant distribution and vegetation development.
- A tundra-type breeding bird assemblage showing general similarity to, but specific differences from, that occurring on arctic–subarctic tundras.
- Significant fractions of the total populations of certain breeding bird species in Europe and particularly in the territories of the European Communities (Table 8.1).
- Insular ecological and other adaptations by several bird species which may represent incipient evolutionary divergence in Britain.

The crucial question remaining is about how much of the total peatland area now left in Caithness and Sutherland deserves national and international conservation designation or other conservation measures. Nature conservation faces a new situation in the Caithness and Sutherland peatlands, in that the distinction usually made elsewhere between specially important sites and the more diffused interest of 'the wider countryside' breaks down here. Over much of Britain, especially the lowlands, a great deal of the land surface has been so modified by Man that nature conservation interest is now thinly spread, so that remnants of semi-natural habitat are readily identifiable as specially important 'islands'. Even in upland areas where semi-natural habitat remains extensive, it may be of variable nature conservation interest, so that selecting the best areas is still possible, though more difficult. On the Caithness and Sutherland peatlands, any attempt to select the 'best' areas for conservation measures is much more problematical and less appropriate, because:

- there are no abrupt differences or

8

Table 8.1 Population and distribution data for birds occurring on the peatlands of Caithness and Sutherland

Species	Arctic breeding species (Sage 1986)	Annex 1 species of EEC Birds Directive	Schedule 1 species of W. & C. Act 1981	Appendix 1 species of Bern Convention	Estimated Caithness/Sutherland peatlands population (pairs)	Estimated British population (pairs)[2]	Percentage of British population in Caithness/Sutherland	Status elsewhere in EC	Percentage of EC population in Caithness/Sutherland	World distribution
Red-throated diver	★	★	★	★	150	1,000–1,200	14%	Absent	14%	Boreal—high arctic
Black-throated diver	★	★	★	★	30	150	20%	Absent	20%	Boreal—mid arctic
Greenland white-fronted goose	★	★			c. 200[3]	c. 9,500[3]	2%	9,300[3] (Ireland)	1%	W. Greenland/Britain/Ireland — restricted and localised
Grey heron		+				3,500–8,500		Scattered		Temperate Palaearctic — at N.W. limit of range
Greylag goose		+	★		c. 300	600–800	43%	Scattered	—[4]	Scattered — eastern continental to subarctic
Wigeon	★	+			80	300–500	20%	Absent	20%	Palaearctic
Teal	★	+				3,500–6,000		Widespread		Widespread — continental to low arctic
Mallard	★	+				40,000+		Widespread		
Common scoter	★	+	★		30+	75–80	39%	c. 100 (Ireland)	16%	Boreal—low arctic
Goldeneye		+	★			>40				
Red-breasted merganser	★	+				1,000–2,000		North only		Boreal—low arctic
Goosander		+				900–1,300				
Hen harrier	★	★	★		30	600	5%	Scattered	1%	Widespread
Sparrowhawk						15–20,000				
Buzzard						8–10,000		Widespread		Widespread
Golden eagle	★	★	★		30	510	6%	Scattered	<1%[5]	Widespread
Kestrel		+		★		30–40,000		Widespread		Widespread
Merlin	★	★	★	★	30	600	5%	Ireland only	4%	Boreal—low arctic — decreasing in numbers in Britain for reasons attributed to land-use change
Peregrine	★	★	★	★	35	730	5%	Scattered	<1%[6]	Widespread
Red grouse								L. l. scoticus elsewhere only in Ireland		Boreal—low arctic — British/Irish race decreasing throughout range
Black grouse						10–50,000				Northern coastal to low arctic
Oystercatcher	★	+				33–43,000		Mainly northern and coastal		Northern coastal to low arctic
Ringed plover	★	+		★		8,600		Scattered — northern coastal		Northern coastal to mid arctic
Golden plover	★	★			3,980	22,600	18%	<650 prs	17%	Boreal—mid arctic, but several distinct races: most of temperate population breeding in Britain
Lapwing		+			500	181,500	<1%	Scattered		Boreal—continental — Britain holding highest numbers in Europe
Temminck's stint	★	+	★	★	<10	<10	—	Absent	—	Montane boreal—low arctic
Dunlin	★	+		★	3,830	9,900	39%	<1,000 prs	35%	Boreal—mid arctic, but temperate population largely restricted to Britain
Snipe	★	+			c. 500+	29,600	3%	Widespread but local		Boreal—low arctic arctic
Woodcock		+				8–35,000		Scattered		Boreal—low arctic: Palaearctic to N. India
Ruff	★	★	★		<10	10–12	—	Local in Low Countries: c. 2,000 prs	—	Temperate—boreal—low arctic
Curlew		+			500	33–38,000	1%	<10,000 prs	1%	Mid continental—subarctic
Redshank		+			100	32,100	<1%	Scattered		Widespread — continental, mainly northern
Greenshank		+	★		630	960	66%	Absent	66%	Boreal to edge of low arctic — mainly natural forest bogs in Fenno-Scandia

Species	Arctic breeding species (Sage 1986)	Annex 1 species of EEC Birds Directive	Schedule 1 species of W. & C. Act 1981	Appendix 1 species of 'Bern' Convention	Estimated Caithness/Sutherland peatlands population (pairs)	Estimated British population (pairs)[2]	Percentage of British population in Caithness/Sutherland	Status elsewhere in EC	Percentage of EC population in Caithness/Sutherland	World distribution
Wood sandpiper	★	★	★	★	<10	1-12	—	c. 150 prs	—	Boreal to edge of low arctic
Common sandpiper		+		★	500	17-20,000	3%	Scattered		Widespread — continental to low arctic
Red-necked phalarope	★	★	★	★	<10	19-24	—	Absent	—	Montane boreal—low arctic
Arctic skua	★	+			60+	2,800+	2%	Absent	2%	Boreal—high arctic
Black-headed gull		+				120-220,000		Widespread but scattered		Continental—boreal
Common gull	★	+			c. 4,000	40,000	10%	Scattered		Northern continental to low arctic
Great black-backed gull		+				22,000		Scattered and localised — Ireland/France/Denmark		Coastal — North Atlantic
Lesser black-backed gull		+				70,000+		Scattered — coastal		Continental coastal to low arctic
Short-eared owl	★	★		★	50	1,000+	5%	Widespread	4%	Widespread
Skylark		+				2 million		Widespread		Widespread
Meadow pipit	★	+				1-1.5 million		Widespread		Widespread—continental—subarctic
Grey wagtail		+		★		15-40,000				
Pied wagtail		+		★		300,000				
Dipper				★		20-25,000		Scattered — subalpine		Widespread — sub-montane—subarctic; declines in highly afforested areas attributed to acid run-off
Whinchat		+		★		15-30,000				
Stonechat		+		★		20-40,000				
Wheatear	★	+		★		60,000		Widespread		Widespread
Ring ouzel		+				6-12,000		Subalpine/alpine		Alpine to boreal—subarctic
Sedge warbler		+		★		200,000				
Hooded crow						1 million				
Raven						4,000				
Twite		+		★		15-30,000				

1 Species marked ★ are listed on Annex 1 of the EEC Directive on the Conservation of Wild Birds as requiring special protection measures, particularly as regards their habitat under Article 4(1). Species marked + are migratory and require similar habitat protection measures under Article 4(2).
2 This excludes the whole of Ireland.
3 Individuals.
4 EC population uncertain owing to unknown proportion of feral birds in other populations. The population in north-west Scotland is the only one thought to be natural, owing to separation from others.
5 Most of the EC population is of the south European race *homeyeri*; Britain holds all of the EC population of the nominate race, 6% of which occur on the Caithness and Sutherland peatlands.
6 Most of the EC population consists of the Mediterranean race *brookei*; Caithness and Sutherland peatlands hold 5% of the EC population of the nominate race.

distinct boundaries in special interest within the peatland areas, other than those recently created by afforestation;
- within any one peatland area, the overall high level of special interest is reinforced by the mutual interdependence of the different habitat and species components of the ecosystem;
- the international importance of these peatlands resides especially in their total extent and wildlife content.

These points will now be amplified, first by considering the needs of the bog structure and vegetation and of the birds separately.

As regards the structural and vegetational characteristics of these peatlands, it is possible to identify undisturbed pool and hummock systems, level flows and associated fens as outstandingly important features. These cannot be considered in isolation and have to be regarded as nuclei within the whole moorland ecosystem which needs to be represented, including the related wet and dry heaths, flushes, lochs and streams. In making any such selection of representative areas, the complete range of variation in surface pattern structure and vegetation and in botanical features illustrating geographical, topographic and altitudinal variation within the region also has to be included. The location and extent of these key peatland areas, all of which qualify as "key sites" in the terms of *A nature conservation review* (Ratcliffe 1977b) are shown in Figure 8.1. The concept of delineating undisturbed catchments has to be expressed in the selection process, but, because many of the most important flows lie across broad and ill-defined watersheds, boundaries drawn through the imaginary middle of watersheds surrounding a particular catchment are quite inappropriate: to protect these features they have to be drawn well within adjacent catchments. Extensive hydrological systems which have less than 6% cover of conifer plantations are shown in Figure 8.2. There is also the problem of the effect of afforestation on ground beyond the forest edge: this extends over a wide zone, but the total effect is not known. Aerial spreading of fertilisers and pesticides is also likely to affect a peripheral zone of unknown width beyond the forest edge.

Ornithological site assessment has to take account of the point that each bird species occupies its own ecological niche and that the total bird assemblage thus requires all the different habitat components of the moorland ecosystem. While the pool and hummock systems are especially important for several species, feeding and nesting habitat requirements usually diverge. Greenshanks feed at pools, lochs and rivers but often nest on shallow peat or dry, stony moraines. Golden plovers, curlews, red grouse, meadow pipits and skylarks nest in large total numbers on the shallow peat areas as well as on the wetter flows. Snipe and redshanks feed in the grassier and flushed areas. The interpolation procedures described and validated in section 4.2 allow areas of high quality wader habitat to be identified, all appropriate for key site status. The distribution of these areas, with a necessary buffer zone of 1 km to allow for the effect of plantations both on the adjacent peatland vegetation and on the moorland bird assemblage, is shown in Figure 8.3.

The other waterfowl need lochs and rivers, and their numbers depend on the frequency and extent of these habitats. Most of the ducks and greylag geese nest on the moorland well back from the edges of these open water habitats. The raptors, owls and scavenging birds are widely dispersed over the whole area, each pair holding a large territory and needing an extensive hunting range (up to 10,000 ha for a pair of golden eagles). Because of the mobility of birds and their need to range widely but variably, the drawing of boundaries through a continuous peatland area becomes an even more arbitrary and unsatisfactory process than it does in trying to delineate representative sites for their structural and vegetational interest.

The possible 'edge-effects' from

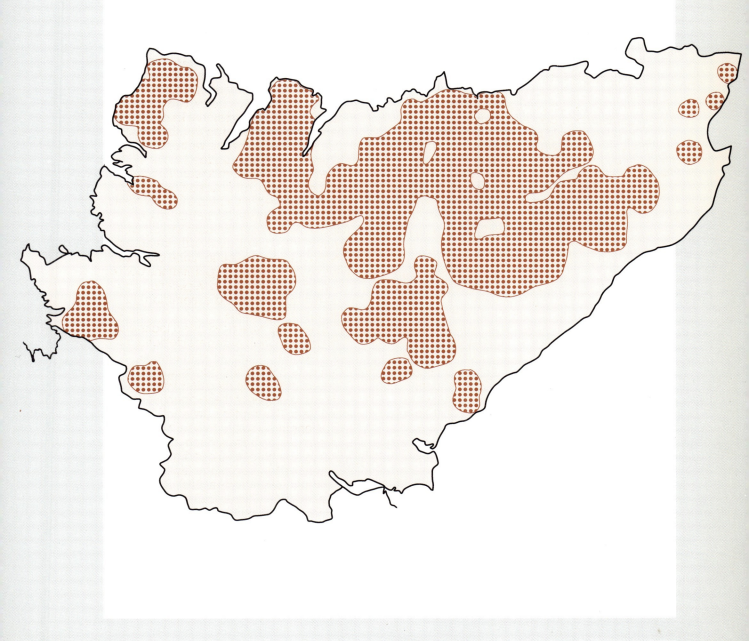

Figure 8.1 Extent of provisional key peatland areas in Caithness and Sutherland including associated hydrological units and 1km buffer zones

Key

▦ Provisional key peatlands

Figure 8.2 Key unafforested hydrological systems in Caithness and Sutherland.

The shaded areas represent 23 individual systems which had less than 6% cover of conifer plantations in January 1986

Key

Key hydrological systems

Figure 8.3 Extent of high quality peatland wader habitat (categories A and B of Table 4.3) in Caithness and Sutherland including associated 1km buffer zones

Key

▥ High quality wader habitat

8

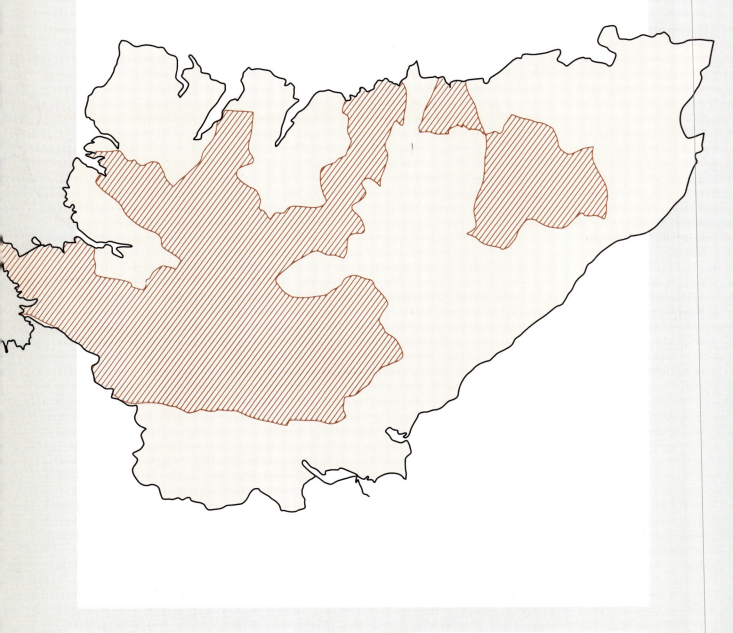

Figure 8.4 River catchments (including loch systems) of exceptional importance for black-throated divers

Key

▨ River catchments

Figure 8.5 Area containing significant known merlin nesting and feeding habitat and nesting sites of rare breeding waders and common scoter up to 1986 (10km grid squares)

Key

 Rare breeding birds habitat

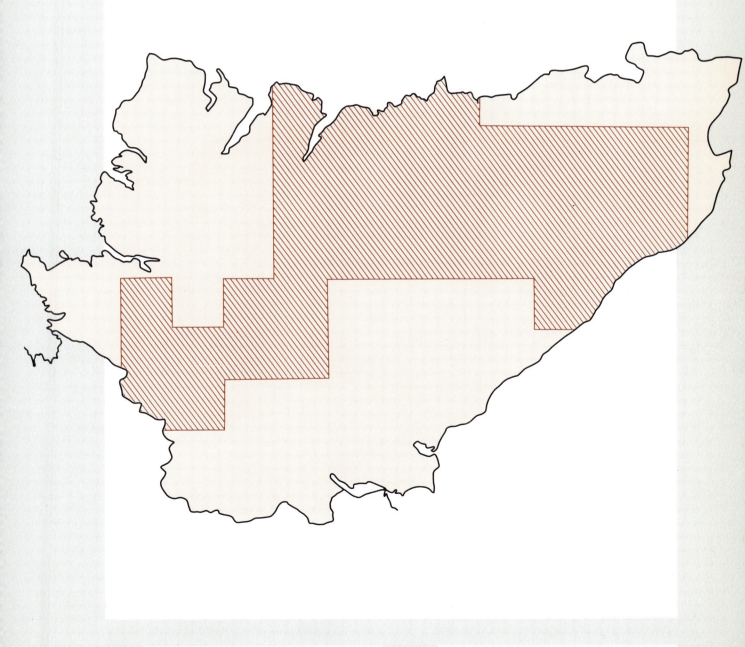

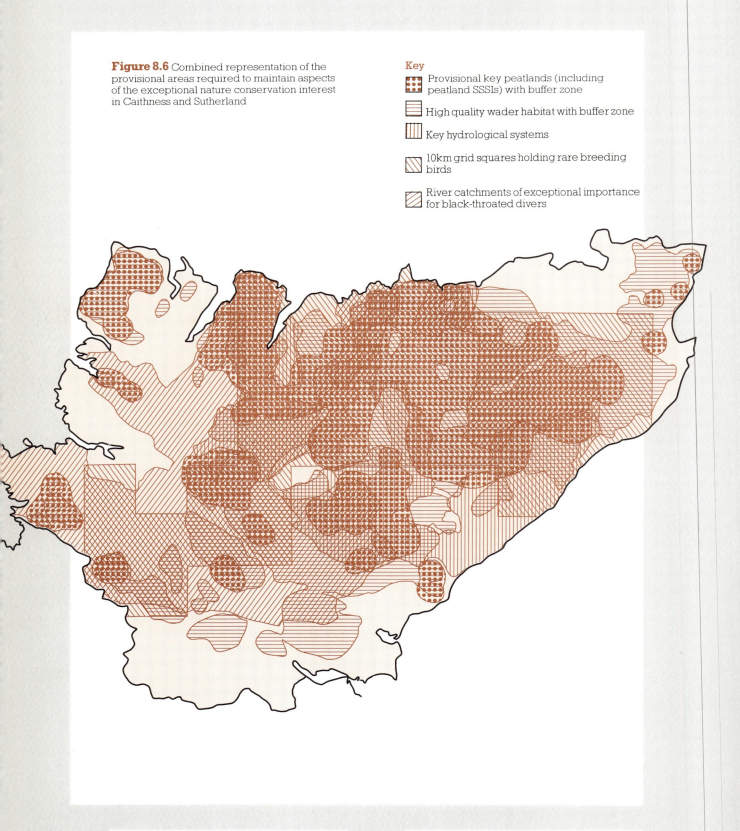

Figure 8.6 Combined representation of the provisional areas required to maintain aspects of the exceptional nature conservation interest in Caithness and Sutherland

Key

- Provisional key peatlands (including peatland SSSIs) with buffer zone
- High quality wader habitat with buffer zone
- Key hydrological systems
- 10km grid squares holding rare breeding birds
- River catchments of exceptional importance for black-throated divers

Figure 8.7 Extent of blanket bog in Caithness and Sutherland, with areas of forestry (including land in Forestry Commission ownership or with Forest Grant Scheme approval) established on peatland and elsewhere

Key
- Blanket bog
- Plantations on blanket bog
- Plantations on other substrates

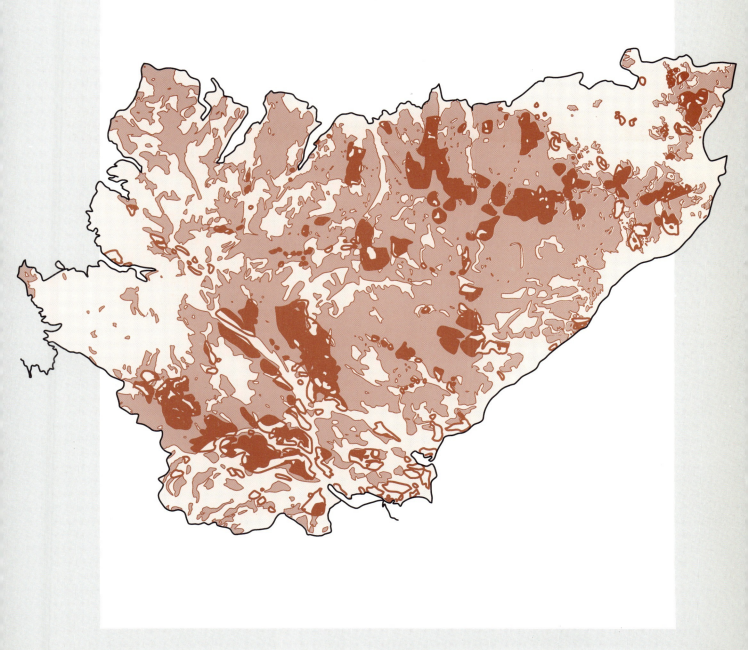

encroaching forest also have worrying portents for breeding birds, and some of them may not show until the plantations are well established. Any avoidance effects are likely to be overtaken in seriousness by increased nest predation from crows and foxes (see section 6.1). The effects of afforestation on hydrology, sedimentation and water chemistry of lochs and rivers, both within and outside the planted ground, could also have adverse implications for edge-feeding waders, other waterfowl and riparian birds. It is important that complete large river catchments are protected from afforestation, since acidification is known to be increased by conifer plantations in base-poor geological areas and adversely affects breeding water birds such as black-throated divers and dippers (section 6.5) and the economically important populations of salmon and trout (section 5.1; see also Nature Conservancy Council 1986, pp. 40-41). Twelve key river catchments each holding more than 2% of the EC population of black-throated divers are shown in Figure 8.4. Quite often, high ornithological interest and high vegetational interest do not coincide. The distribution of the rarer birds (Figure 8.5) and plants is erratic and uncorrelated.

These considerations underline the great difficulty in trying to delineate arbitrary units of peatland which could satisfy the need to represent adequately the total field of variation and to ensure that the units could be individually viable and secure from gradual loss of interest. For it has to be assumed that most of the surrounding areas would be vulnerable to further afforestation, right up to the boundaries. While there is a general correlation in quality between certain structural/vegetational and ornithological features, an attempt to select a representative series of sites for the one would by no means take adequate account of the other. Figure 8.6 combines the areas required to maintain the features described above. It is clear that the exceptional interest is maintained throughout much of the peatlands and that any one part of them is of importance to several overlapping interests.

Beyond this, however, the overriding issue is that the concept of a representative series of exemplary sites is no longer accepted as an adequate recognition of the conservation value and needs of the Caithness and Sutherland peatlands. This earlier approach does not match the present reassessment, based on fuller survey information and the international dimension. The outstanding international importance of these peatlands lies in their total extent, continuity and diversity as mire forms and vegetation complexes and in the total size and species composition of their bird populations.

The remaining extent of the Caithness and Sutherland peatlands as a whole should therefore be regarded as the desirable nature conservation area for its national and international importance. A significant and important fraction of the total has already been lost to afforestation in an extremely arbitrary and haphazard way (Figure 8.7), and there is no rational scientific or conservation basis for making a further arbitrary selection within the remainder, to surrender additional areas to afforestation. The case for retaining as much as possible of what remains is reinforced by the extensive losses of peatlands and their wildlife resulting from afforestation in other parts of Britain (Wales, the Cheviots, the Southern Uplands of Scotland and the Highlands and Islands) and by the fact that such losses will continue elsewhere under a policy which sets an open-ended annual planting target of 33,000 ha. Indeed, under the present rules which promote afforestation, all plantable areas of blanket bog could eventually disappear outside specially protected areas.

In the time that it has taken to collect and analyse the data presented in this report (1979-1986), habitat supporting nearly 19% of the specialised breeding wader populations has been destroyed and only eight of 41 hydrological systems in Caithness and eastern Sutherland have been left free of afforestation.

The maintenance of the nature conservation interest over the peatlands is largely compatible with the traditional land-uses, which include crofting, game management and fishing. In contrast to these traditional uses of peatland areas, the drastic change in land-use caused by afforestation is totally destructive to the natural and semi-natural habitats and their nature conservation value. While crofting, game management and nature conservation are not mutually exclusive, blanket afforestation also forecloses future options on traditional forms of land-use as effectively as it damages nature conservation interests. It thus follows that the simplest way of achieving nature conservation over the Caithness and Sutherland peatlands is to maintain existing land-uses, though with greater regulation of moor-burning. Beyond making this point, the present report does not seek to discuss details of the possible conservation measures which might be adopted to ensure this outcome, nor is it concerned with the questions that have been raised concerning the economic and social justification of such forestry (National Audit Office 1986). Its aim is to present the conservation case.

Dubh lochan systems, central Caithness

8

During recent years, a great deal of concern has been expressed over the past losses of natural and semi-natural habitat and its wildlife in Britain. Many of these losses took place during earlier periods before nature conservation was conceived — notably the destruction of the great forests and the fenlands before 1800 AD. The post-1940 inroads into the coastlands, chalk grasslands, lowland heaths, moorlands, old hay meadows, marshes and hedges and the pollution of lakes and rivers mostly occurred in response to national need or before effective legislation and adequate knowledge existed. The area of the Caithness and Sutherland peatlands already lost to forestry — most of it since the passing of the Wildlife and Countryside Act 1981 — represents perhaps the most massive single loss of important wildlife habitat since the Second World War. Henceforth the situation will be different, in that any further losses to this unique area will take place as the result of deliberate decisions taken in full knowledge of what is at stake.

Acknowledgements 9

The Upland Bird Survey was funded by the Nature Conservancy Council's Chief Scientist Directorate (Project No. 406). Work in 1985 and 1986 formed part of NCC's Moorland Bird Study (Project No. 433). We are most grateful to all the owners and occupiers of survey plots for permission to work on their land and to their gamekeepers and agents for local help and advice.

We record our thanks to Dr Derek Langslow for initiating the survey and to Fraser Symonds for developing the techniques in the pilot years.

The fieldwork was undertaken by Dr Tim Reed, David Stroud, Fraser Symonds, Chris McCarty, John Barrett, Catrina Barrett, Martin Moss, Kevin Shepherd and Phil Ball with assistance from Dr Derek Langslow. Work used to test methodology for this programme was undertaken by NCC and RSPB staff including Robert Barton, Roger Buisson, Dr Tony Fox, Pete Hack and Chris Thomas, with additional data from Dr Peter Ferns and Roger Lovegrove. Ian Mitchell kindly extracted information on recent afforestation from the Forestry Commission's stock maps in Inverness. We wish to acknowledge the help given to the survey teams by NCC's Regional Officer for North-West Scotland, Dr Peter Tilbrook, and Assistant Regional Officers for Caithness and Sutherland, Stuart Angus, Lesley Cranna, Terry Keatinge and Kristin Scott. Their advice and assistance over the years in ironing out problems has been invaluable.

NCC's peatland surveys have been undertaken by Fiona Burd, Mandy Camm, Dan Charman, Fiona Everingham, Sarah Garnett, Richard Lindsay, Sara Oldfield, Rachel O'Reilly, John Ratcliffe, John Riggall, Jane Smart, David Stroud, Colin Wells and Sylvia White. Fiona Everingham also gave help in drafting this report.

We are grateful to RSPB and Dr Len Campbell for allowing us to use the results of their Caithness and Sutherland survey of 1985 and other unpublished data. Their survey was undertaken by M. J. Birkin and S. J. Hayhow. Roy Dennis and Dr Greg Mudge helpfully supplied information on the distribution of rare bird species. Dr Derek Ratcliffe and Dr Des Thompson of NCC provided valuable additional information on the distribution of waders and raptors in Caithness and Sutherland, and Margaret Palmer and Dr Ian McLean gave helpful advice on freshwater and invertebrate matters.

Dr Judy Stroud helped with mapping land-uses, Fiona Burd categorised maps of landforms, and earlier drafts of the report were read by I. S. Angus, Dr M. I. Avery, Dr I. Bainbridge, Dr C. Bibby, Dr H. J. B. Birks, Dr L. Campbell, Dr P. J. Dugan, Dr N. Easterbee, Dr P. R. Evans, Dr A. D. Fox, Dr R. J. Fuller, Dr H. Galbraith, Dr M. W. Holdgate, Dr T. H. Keatinge, Dr A. N. Lance, Dr D. R. Langslow, R. G. Soutar and Mrs B. W. Vittery. We are particularly grateful to Dr D. A. Ratcliffe, who made numerous and substantial contributions to the text.

Special thanks are due to Kevin Shepherd, who carried out analyses, and particularly to Alison Rothwell, who gave great help in the production of this report, proof-reading several drafts and preparing most of the figures.

We should also like to express our deep gratitude to Sheila Beech, Korina Brighton, Janet Holding and Maureen Sladden (Word Processor Operators at Peterborough) for typing many draft versions of this report and struggling successfully with the large tables, to Marina Palmer for aiding production, to Barbara Brown for clerical assistance, and to Stuart Wallace, who helped in preparing maps and diagrams.

We thank Doric Computer Systems for the help and expertise provided to us during the production of the computer generated maps contained herein. The Arc/Info Computer Mapping/Graphic Information System was used to process the many classes of data needed for the study and to generate the maps and analyses.

Finally, we record our warmest thanks to Philip Oswald for editing the text for publication.

References

ANGUS, S. 1987.
The peatland habitat resource of Sutherland and Caithness in relation to forestry. Unpublished NCC report.

BARRETT, J., REED, T. M., BARRETT, C., & LANGSLOW, D. R. in prep.
Breeding waders of the Sutherland blanket bogs.

BELLAMY, D. 1986.
Bellamy's Ireland: the wild boglands. Dublin, Country House.

BIBBY, C. J., & NATTRASS, M. 1986.
Breeding status of the Merlin in Britain. *British Birds, 79,* 170-185.

BIRKIN, M. J., HAYHOW, S. J., & CAMPBELL, L. H. 1985.
Moorland breeding bird survey. Caithness and Sutherland 1985. Unpublished Royal Society for the Protection of Birds report.

BIRKS, H. H. 1975.
Studies in the vegetational history of Scotland. IV. Pine stumps in Scottish blanket peats. *Royal Society. Philosophical Transactions (B), 270,* 181-226.

BIRKS, H. H. 1984.
Late Quaternary pollen and plant macrofossil stratigraphy at Lochan an Druim, north-west Scotland. *In: Lake sediments and environmental history. Studies in palaeolimnology and palaeoecology in honour of Winifred Tutin,* ed. by E. Y. Haworth and J. W. G. Lund, 377-405. Leicester University Press.

BIRKS, H. J. B., & RATCLIFFE, D. A. 1980.
Upland vegetation types. A list of National Vegetation Classification plant communities. Unpublished NCC report.

BOATMAN, D. J., & ARMSTRONG, W. 1968.
A bog type in north-west Sutherland. *Journal of Ecology, 56,* 129-141.

BRAGG, O., LINDSAY, R. A., ROBERTSON, H., & HEATON, A. in prep.
A historical review of lowland raised mires. Peterborough, Nature Conservancy Council.

BROAD, R. A., SEDDON, A. J. E., & STROUD, D. A. 1986.
The waterfowl of freshwater lochs in Argyll: May-June 1985. *Argyll Bird Report, 13,* 77-88.

BUNDY, G. 1979.
Breeding and feeding observations on the Black-throated Diver. *Bird Study, 26,* 33-36.

BYRNE, R. W., & MACKENZIE-GRIEVE, C. J. 1974.
Pectoral Sandpipers in Caithness and Shetland. *Scottish Birds, 8,* 72-73

CALDER, I. R., & NEWSON, M. D. 1979.
Land use and upland water resources in Britain — a strategic look. *Water Resources Bulletin, 15,* 1628-1639.

CAMPBELL, L. H. 1985.
Habitat features as a means of identifying areas of importance for moorland breeding birds. *In: Bird census and atlas studies. Proceedings of the VIII International Conference on Bird Census and Atlas Work,* ed. by K. Taylor, R. J. Fuller and P. C. Lack, 269-272. Tring, British Trust for Ornithology.

CAMPBELL, L. H., & TALBOT, T. R. 1987.
The breeding status of Black-throated Diver *(Gavia arctica)* in Scotland. *British Birds, 80,* 1-8.

CHAPMAN, S. B., & ROSE, R. J. 1986.
An assessment of changes in the vegetation at Coom Rigg Moss National Nature Reserve within the period 1958-1986. Unpublished report from Natural Environment Research Council (Institute of Terrestrial Ecology) to Nature Conservancy Council.

CHARMAN, D. J. 1986.
The influence of area, habitat diversity and isolation period on species numbers of the Border Mires of Northumberland. B. Sc. thesis, University of Newcastle upon Tyne.

CLYMO, R. S. 1983.
Peat. *In: Mires: Swamp, bog, fen and moor. General studies,* ed. by A. J. P. Gore, 159-224. Amsterdam, Elsevier (Ecosystems of the World 4A).

CRAMP, S., & SIMMONS, K. E. L., eds. 1977.
Handbook of the birds of Europe, the Middle East and North Africa. The birds of the Western Palearctic. Volume I. Oxford University Press.

CRAMP, S., & SIMMONS, K. E. L., eds. 1980.
Handbook of the birds of Europe, the Middle East and North Africa. The birds of the Western Palearctic. Volume II. Oxford University Press.

CRAMP, S., & SIMMONS, K. E. L., eds. 1983.
Handbook of the birds of Europe, the Middle East and North Africa. The birds of the Western Palearctic. Volume III. Oxford University Press.

CRAMP, S., & SIMMONS, K. E. L., eds. 1985. *Handbook of the birds of Europe, the Middle East and North Africa. The birds of the Western Palearctic. Volume IV.* Oxford University Press.

CRAMPTON, C. B. 1911. *The vegetation of Caithness considered in relation to the geology.* Committee for the Survey and Study of British Vegetation.

CURTIS, D. J., & BIGNAL, E. M. 1985. Quantitative description of vegetation physiognomy using vertical quadrats. *Vegetatio, 63,* 97-104.

DENNIS, R. H. 1976. Scottish bird report 1975. *Scottish Birds, 9,* 173-235.

DENNIS, R. H., & DOW, H. 1984. The establishment of a population of Goldeneye *Bucephala clangula* breeding in Scotland. *Bird Study, 31,* 217-222.

DENNIS, R. H., ELLIS, P. M., BROAD, R. A., & LANGSLOW, D. R. 1984. The status of the Golden Eagle in Britain. *British Birds, 77,* 592-607.

DRAKEFORD, T. 1979. *Report of survey of the afforested spawning grounds of the Fleet Catchment.* Dumfries, Forestry Commission (South Scotland Conservancy) (unpublished report).

DRAKEFORD, T. 1982. Management of upland streams (an experimental fisheries management project on the afforested headwaters of the River Fleet, Kirkcudbrightshire). *Institute of Fisheries Management. 12th Annual Study Course, Durham,* 86-92.

EGGLISHAW, H., GARDINER, R., & FOSTER, J. 1986. Salmon catch decline and forestry in Scotland. *Scottish Geographical Magazine, 102,* 57-61.

EINARSSON, T. 1968. On the formation and history of Icelandic peat bogs. *Proceedings of Second International Peat Congress, 1,* 213-215.

ERIKSSON, M. O. 1984. Acidification of lakes: effects on waterbirds in Sweden. *Ambio, 13,* 260-262.

ERIKSSON, M. O. G., HENRIKSON, L., & OSCARSON, M. G. 1980. Sjofagel och forsurning-nagra Synpunkter. *Vår Fågelvärld, 39,* 163-166.

EUROLA, S., & HOLAPPA, K. 1985. The Finnish mire type system. *Aquilo Ser. Botanica, 21,* 101-110.

EVERETT, M. J. 1982. Breeding Great and Arctic Skuas in Scotland in 1974-75. *Seabird Report, 6,* 50-58.

FORESTRY COMMISSION. 1957. *Exotic forest trees in Great Britain.* (Forestry Commission Bulletin No. 30).

FOSTER, G. N. 1987. *Surveys of three rare water beetles.* Unpublished report to Nature Conservancy Council.

FOSTER, G. N., & SPIRIT, M. 1986. *Oreodytes alpinus* new to Britain. *Balfour-Browne Club Newsletter, 36,* 1-2.

FOX, A. D. 1986a. The breeding Teal *(Anas crecca)* of a coastal raised mire in central west Wales. *Bird Study, 33,* 18-23.

FOX, A. D. 1986b. Effects of ditch-blockage on adult Odonata at a coastal raised mire site in central west Wales, United Kingdom. *Odonatologica, 15,* 327-334.

FRANCIS, I. S. 1987. *Blanket peat erosion in mid-Wales: two catchment studies.* Ph. D. thesis, University College of Wales, Aberystwyth.

FRY, G. L. A., & COOKE, A. S. 1984. *Acid deposition and its implications for nature conservation in Britain.* Shrewsbury, Nature Conservancy Council (Focus on Nature Conservation No. 7).

FULLER, R. J. 1982. *Bird habitats in Britain.* Calton, T. and A. D. Poyser.

FULLER, R. J. 1985. *Studies on breeding waders in the Southern Isles of the Outer Hebrides, 1985.* Report on behalf of Wader Study Group to Nature Conservancy Council (CSD Research Report No. 606).

FULLER, R. J., GREEN, G. H., & PIENKOWSKI, M. W. 1983. Field observations on methods used to count waders breeding at high density in the Outer Hebrides, Scotland. *Wader Study Group Bulletin, 39,* 27-29.

FULLER, R. J., & PERCIVAL, S. M. 1986.
Surveys on breeding waders in the Southern Isles of the Outer Hebrides, 1986.
Report on behalf of Wader Study Group to Nature Conservancy Council (CSD Research Report No. 690).

FURNESS, R. W. 1986.
The conservation of Arctic and Great Skuas and their impact on agriculture.
Unpublished report to Nature Conservancy Council.

GARFITT, J. E. 1983.
Afforestation and upland birds. *Quarterly Journal of Forestry, 77,* 253-254.

GODWIN, H. 1975.
The history of the British flora. 2nd ed.
Cambridge University Press.

GOMERSALL, C. H., MORTON, J. S., & WYNDE, R. M. 1984.
Status of breeding Red-throated Divers in Shetlands, 1983. *Bird Study, 31,* 223-229.

GOODE, D. A. 1973.
The significance of physical hydrology in the morphological classification of mires. *In: Proceedings of the International Peat Society, Glasgow. Classification of peat and peatlands. September 1973.*

GOODWILLIE, R. 1980.
European peatlands. Strasbourg, Council of Europe (Nature and Environment Series, No. 19).

GORDON, J. E. 1981.
Ice-scoured topography and its relationships to bedrock structure and ice-movement in parts of northern Scotland and West Greenland. *Geografiska Annaler, 63A,* 55-65.

GORE, A. J. P., ed. 1983.
Mires: Swamp, bog, fen and moor. Regional studies. Amsterdam, Elsevier (Ecosystems of the World 4B).

GRAESSER, N. W. 1979.
Effects on salmon fisheries of afforestation, land drainage and road making in river catchments areas. *Salmon Net, 12,* 38-45.

GREEN, R. E. 1985.
Estimating the abundance of breeding Snipe. *Bird Study, 32,* 141-149.

GREENLAND WHITE-FRONTED GOOSE STUDY. 1986.
Greenland White-fronted Geese in Britain; 1985-86. Aberystwyth.

HAKALA, A. 1971.
A quantitative study of the bird fauna of some open peatlands in Finland. *Ornis Fennica, 48,* 1-11.

HARRIMAN, R., & MORRISON, B. R. S. 1982.
Ecology of streams draining forested and non-forested catchments in an area of central Scotland subject to acid precipitation. *Hydrobiologia, 88,* 251-263.

HARRIS, J. A. 1983.
Birds and coniferous plantations. Royal Forestry Society of England, Wales and Northern Ireland.

HARVIE-BROWN, J. A., & BUCKLEY, T. E. 1887.
A vertebrate fauna of Sutherland, Caithness and West Cromarty. Edinburgh, Douglas.

HARVIE-BROWN, J. A., & BUCKLEY, T. E. 1895.
A fauna of the Moray Basin. 2 vols.
Edinburgh, Douglas.

HEWSON, R., & LEITCH, A. F. 1983.
The food of foxes in forests and on the open hill. *Scottish Forestry, 37,* 39-50.

HOLLAND, P. K., ROBSON, J. E., & YALDEN, D. W. 1982.
The breeding biology of the Common Sandpiper *Actitis hypoleucos* in the Peak District. *Bird Study, 29,* 99-110.

HOLMES, R. T. 1966.
Feeding ecology of the Red-backed Sandpiper *(Calidris alpina)* in arctic Alaska. *Ecology, 47,* 32-45.

HOLMES, R. T. 1970.
Differences in population density, territoriality, and food supply of dunlin on arctic and subarctic tundra. *In: Animal populations in relation to their food resources,* ed. by A. Watson, 303-317. Oxford, Blackwell Scientific for British Ecological Society.

INGRAM, H. A. P. 1982.
Size and shape in raised mire ecosystems: a geophysical model. *Nature, 297* (5864), 300-303.

INTERNATIONAL MIRE CONSERVATION GROUP. 1986.
Press release. 30 September 1986.

INTERNATIONAL UNION FOR CONSERVATION OF NATURE AND NATURAL RESOURCES. 1984.
Convention on Wetlands of International Importance especially as Waterfowl Habitat. Proceedings of the Second Conference of the Contracting Parties. Gland, Switzerland.

IUCN/UNEP/WWF. 1980.
World Conservation Strategy. Living resource conservation for sustainable development. Gland, Switzerland.

IVANOV, K. E. 1981.
Water movement in mirelands. London, Academic Press.

JACKSON, D. B., & PERCIVAL, S. M. 1983.
The breeding waders of the Hebridean machair: a validation check of the census methods. *Wader Study Group Bulletin, 39,* 20-24.

JOHANSEN, J. 1975.
Pollen diagrams from the Shetland and Faroe Islands. *New Phytologist, 75,* 369.

LANGSLOW, D. R. 1983.
The impacts of afforestation on breeding shorebirds and raptors in the uplands of Britain. *In: Shorebirds and large waterbirds conservation,* ed. by P. R. Evans, H. Hafner and P. L'Hermite, 17-25. Brussels, Commission of the European Communities.

LANGSLOW, D. R., & REED, T. M. 1985.
Inter-year comparisons of breeding wader populations in the uplands of England and Scotland. *In: Bird census and atlas studies. Proceedings of the VIII International Conference on Bird Census and Atlas Work,* ed. by K. Taylor, R. J. Fuller and P. C. Lack, 165-173. Tring, British Trust for Ornithology.

LAYBOURNE, S., & FOX, A. D. in press.
Greenland White-fronted Geese in Caithness. *Scottish Birds.*

LAYBOURNE, S., MANSON, S. A. M., & COLLETT, P. M. 1977.
Arctic terns breeding inland in Caithness. *Scottish Birds, 9,* 301.

LINDSAY, R. A. 1987.
The great flow — an international responsibility. *New Scientist, 1542* (8 January), 45.

LINDSAY, R. A. in prep.
Hydrophysics of drainage and ecological consequences on ombrotrophic mires. Peterborough, Nature Conservancy Council.

LINDSAY, R. A., *et al.* in prep.
Peatlands of Caithness and Sutherland. Peterborough, Nature Conservancy Council.

LINDSAY, R. A., RIGGALL, J., & BIGNAL, E. M. 1983.
Ombrogenous mires in Islay and Mull. *Royal Society of Edinburgh. Proceedings, 83B,* 341-371.

LINDSAY, R. A., RIGGALL, J., & BURD, F. H. 1985.
The use of small-scale surface patterns in the classification of British peatlands. *Aquilo, 21,* 69-79.

McDONALD, A. 1973.
Some views on the effects of peat drainage. *Scottish Forestry, 27,* 315-327.

McVEAN, D. N., & RATCLIFFE, D. A. 1962.
Plant communities of the Scottish Highlands. Nature Conservancy Monograph No. 1. London, HMSO.

MARQUISS, M., NEWTON, I., & RATCLIFFE, D. A. 1978.
The decline of the Raven, *Corvus corax,* in relation to afforestation in southern Scotland and northern England. *Journal of Applied Ecology, 15,* 129-144.

MARQUISS, M., RATCLIFFE, D. A., & ROXBURGH, L. R. 1985.
The numbers, breeding success and diet of Golden Eagles in Southern Scotland in relation to changes in land use. *Biological Conservation, 34,* 121-140.

MEARNS, R. 1983.
The status of the Raven in Southern Scotland and Northumbria. *Scottish Birds, 12,* 211-218.

MOAR, N. T. 1969.
Two pollen diagrams from the Mainland, Orkney Islands. *New Phytologist, 68,* 201-208.

MOEN, A. 1985.
Classification of mires for conservation purposes in Norway. *Aquilo, 21,* 95-100.

MOORE, P. 1987.
A thousand years of death. *New Scientist, 1542* (8 January), 46-48.

MOSS, D. 1986.
Rain, breeding success and distribution of Capercaillie *Tetrao urogallus* and Black Grouse *Tetrao tetrix* in Scotland. *Ibis, 128,* 65-72.

NATIONAL AUDIT OFFICE. 1986.
Review of Forestry Commission objectives and achievements. London, HMSO.

NATURE CONSERVANCY COUNCIL. 1982.
The conservation of peat bogs. Shrewsbury.

NATURE CONSERVANCY COUNCIL. 1986.
Nature conservation and afforestation in Britain. Peterborough.

NETHERSOLE-THOMPSON, D. & M. 1979.
Greenshanks. Berkhamsted, T. and A. D. Poyser.

NETHERSOLE-THOMPSON, D. & M. 1986.
Waders. Their breeding, haunts and watchers. Calton, T. and A. D. Poyser.

NETHERSOLE-THOMPSON, D., & WATSON, A. 1974.
The Cairngorms. Their natural history and scenery. London, Collins.

NEWTON, I. 1984.
Raptors in Britain — a review of the last 150 years. *BTO News, 131,* 6-7.

NEWTON, I., DAVIS, P. E., & DAVIS, J. E. 1982.
Ravens and buzzards in relation to sheep-farming and forestry in Wales. *Journal of Applied Ecology, 19,* 681-706.

ORMEROD, S. J. 1985.
The diet of breeding Dippers *Cinclus cinclus* and their nestlings in a catchment of the River Wye, mid-Wales: a preliminary study by faecal analysis. *Ibis, 127,* 316-331.

ORMEROD, S. J., BOILSTONE, M. A., & TYLER, S. J. 1985.
Factors influencing the abundance of breeding Dippers *Cinclus cinclus* in the catchment of the River Wye, mid-Wales. *Ibis, 127,* 332-340.

ORMEROD, S. J., TYLER, S. J., & LEWIS, J. M. S. 1985.
Is the breeding distribution of Dippers influenced by stream acidity? *Bird Study, 32,* 32-39.

OWEN, M., ATKINSON-WILLES, G. L., & SALMON, D. 1986.
Wildfowl in Great Britain. 2nd ed. Cambridge University Press.

PARR, W. 1984.
Consultation or confrontation? A review of forestry activities and developments on water catchment areas in south-west Scotland. Institution of Water Engineers and Scientists. Scientific Section. 24 October 1984.

PEARSALL, W. H. 1938.
The soil complex in relation to plant communities. III. Moorlands and bogs. *Journal of Ecology, 26,* 298-315.

PEARSALL, W. H. 1956.
Two blanket bogs in Sutherland. *Journal of Ecology, 44,* 493-516.

PEGLAR, S. 1979.
A radiocarbon-dated pollen diagram from Loch of Winless, Caithness, north-east Scotland. *New Phytologist, 82,* 245-263.

PETTY, S. J. 1985.
Counts of some breeding birds in two recently afforested areas of Kintyre. *Scottish Birds, 13,* 258-262.

PIERSMA, T., ed. 1986.
Breeding waders in Europe: a review of population size estimates and a bibliography of information sources. *Wader Study Group Bulletin, 48, Supplement.*

PRUS-CHACINSKI, T. M. 1962.
Shrinkage of peat-lands due to drainage operations. *Journal of Institute of Water Engineers, 16,* 436-448.

PYATT, D. G. 1987.
Afforestation of blanket peatland — soil effects. *Forestry and British Timber, March,* 15-16.

PYATT, D. G., & CRAVEN, M. M. 1979.
Soil changes under even-aged plantations. *In: The ecology of even-aged forest plantations,* ed. by E. D. Ford, D. C. Malcolm and J. Atterson, 369-386. Cambridge, Institute of Terrestrial Ecology.

RANKIN, G. D., & TAYLOR, I. R. 1985.
Changes within afforested but unplanted ground: birds. Unpublished report to Nature Conservancy Council.

RANKIN, N. 1947.
Haunts of British divers. Collins, London.

RARE BREEDING BIRDS PANEL. 1986.
Rare breeding birds in the United Kingdom in 1984. *British Birds, 79,* 470-495.

RATCLIFFE, D. A. 1968.
An ecological account of Atlantic bryophytes in the British Isles. *New Phytologist, 67,* 365-439.

RATCLIFFE, D. A. 1976.
Observations on the breeding of the Golden Plover in Great Britain. *Bird Study, 23,* 63-116.

RATCLIFFE, D. A. 1977a.
Uplands and birds — an outline. *Bird Study,* *24,* 140-158.

RATCLIFFE, D. A., ed. 1977b.
A nature conservation review. 2 vols. Cambridge University Press.

RATCLIFFE, D. A. 1980.
The peregrine falcon. Calton, T. and A. D. Poyser.

RATCLIFFE, D. A. 1984.
The Peregrine breeding population of the United Kingdom in 1981. *Bird Study, 31,* 1-18.

RATCLIFFE, D. A. 1986.
The effects of afforestation on the wildlife of open habitats. *In: Trees and wildlife in the Scottish uplands,* ed. by D. Jenkins, 46-54. Banchory, Institute of Terrestrial Ecology (ITE Symposium No. 17).

REED, T. M. 1982a.
Birds and afforestation. *Ecos, 3,* 8-10.

REED, T. M. 1982b.
Transect methods. Unpublished NCC report.

REED, T. M. 1985.
Grouse moors and breeding waders. *Game Conservancy Annual Report, 16,* 57-60.

REED, T. M. in prep.
Within-season variation in the detection of Redshank on upland moorlands. *Ibis.*

REED, T. M., BARRETT, C. F., BARRETT, J. C., MAYHOW, S., & MINSHULL, B. 1985.
Diurnal variability in the detection of waders on their breeding grounds. *Bird Study, 32,* 71-74.

REED, T. M., BARRETT, J. C., BARRETT, C., & LANGSLOW, D. R. 1983.
Diurnal variability in the detection of Dunlin *Calidris alpina. Bird Study, 30,* 244-246.

REED, T. M., & FULLER, R. J. 1983.
Methods used to assess populations of breeding waders on machair in the Outer Hebrides. *Wader Study Group Bulletin, 39,* 14-16.

REED, T. M., & LANGSLOW, D. R. 1985.
The timing of breeding in the Golden Plover *Pluvialis apricaria. In: Bird census and atlas studies. Proceedings of the VIII International Conference on Bird Census and Atlas Work,* ed. by K. Taylor, R. J. Fuller and P. C. Lack, 123-129. Tring, British Trust for Ornithology.

REED, T. M., & LANGSLOW, D. R. in press.
Habitat association of breeding waders. *Acta Seria Biologica.*

REED, T. M., & LANGSLOW, D. R. in prep.
The number of breeding bird species on peatlands in southern Sutherland, Scotland. *Biological Conservation.*

REED, T. M., LANGSLOW, D. R., & SYMONDS, F. L. 1983a.
Breeding waders of the Caithness flows. *Scottish Birds, 12,* 180-186.

REED, T. M., LANGSLOW, D. R., & SYMONDS, F. L. 1983b.
Arctic Skuas in Caithness 1979 and 1980. *Bird Study, 30,* 24-26.

REED, T. M., WILLIAMS, T. D., & WEBB, A. 1983.
The Wader Study Group survey of Hebridean waders: was the timing right? *Wader Study Group Bulletin, 39,* 17-19.

REYNOLDS, J. 1984.
Vanishing Irish boglands. *World Wildlife News, Spring 1984,* 10-16.

ROBINSON, M. 1980.
The effect of pre-afforestation drainage on the streamflow and water quality of a small upland catchment. Wallingford, Institute of Hydrology (Report No. 73).

ROBINSON, M. 1985.
The hydrological effects of moorland gripping: a re-appraisal of the Moor House research. *Journal of Environmental Management, 21,* 205-211.

ROBINSON, M., & BLYTH, K. 1982.
The effect of forestry drainage operations on upland sediment yields: a case study. *Earth Surface Processes and Landforms, 7,* 85-90.

ROYAL SOCIETY FOR THE PROTECTION OF BIRDS. 1985.
Forestry in the flow country — the threat to birds. Sandy.

RYAN, J. B., & CROSS, J. R. 1984.
The conservation of peatlands in Ireland. *Proceedings of the International Peat Congress, Dublin,* 388-406.

SAGE, B. 1986.
The Arctic and its wildlife. Beckenham, Croom Helm.

SAMMALISTO, L. 1957.
The effect of the woodland—open peatland edge on some peatland birds in South Finland. *Ornis Fennica, 34,* 81-89.

SCOTTISH PEAT COMMITTEE. 1968.
Scottish peat surveys. Volume 4. Caithness, Shetland and Orkney. Department of Agriculture and Fisheries for Scotland. Edinburgh, HMSO.

SCOTTISH WILDLIFE TRUST. 1987.
The future of the flows. Edinburgh.

SHARROCK, J. T. R., ed. 1976.
The atlas of breeding birds in Britain and Ireland. Tring, British Trust for Ornithology/Irish Wildbird Conservancy.

SILVOLA, J. 1986.
Carbon dioxide dynamics in mires reclaimed for forestry in eastern Finland. *Annales Botanici Fennici, 23,* 59-67.

SOIKKELI, M. 1970.
Dispersal of Dunlin *Calidris alpina* in relation to sites of birth and breeding. *Ornis Fennica, 47,* 1-9.

SPIRIT, M. 1986.
Biological recording in Caithness. *Balfour-Browne Club Newsletter, 37,* 16-17.

STEWART, P. J. 1987.
Growing against the grain. United Kingdom forestry policy, 1987. Council for the Protection of Rural England.

STONER, J. H., GEE, A. S., & WADE, K. R. 1984.
The effects of acidification on the ecology of streams in the upper Tywi catchment in west Wales. *Environmental Pollution, A, 35,* 125-157.

STONER, J. H., & GEE, A. S. 1985.
Effects of forestry on water quality and fish in Welsh rivers and lakes. *Institute of Water Engineers and Scientists. Journal, 39,* 27-45.

STROUD, D. A. 1985.
Greenland White-fronted Geese in Britain; 1984-85. Aberystwyth, Greenland White-fronted Goose Study.

STROUD, D. A., & REED, T. M. 1986.
The effect of plantation proximity on moorland breeding waders. *Wader Study Group Bulletin, 46,* 25-28.

SYMONDS, F. L. 1981.
A survey of breeding waders and wildfowl in Caithness. *Wader Study Group Bulletin, 31,* 9.

TANSLEY, A. G. 1939.
The British Islands and their vegetation. Cambridge University Press.

THOM, V. M. 1986.
Birds in Scotland. Calton, T. and A. D. Poyser.

THOMPSON, D. A. 1979.
Forest drainage schemes. Forestry Commission (Leaflet No. 72).

THOMPSON, D. A. 1984.
Ploughing of forest soils. Forestry Commission (Leaflet No. 71).

THOMPSON, D. B. A. 1987.
Battle of the bog. *New Scientist, 1542* (8 January), 41-44.

THOMPSON, D. B. A., THOMPSON, P. S., & NETHERSOLE-THOMPSON, D. 1986.
Timing of breeding and breeding performance in a population of greenshanks *Tringa nebularia. Journal of Animal Ecology, 55,* 181-199.

THOMPSON, D. B. A., THOMPSON, P. S., & NETHERSOLE-THOMPSON, D. in press.
Breeding site fidelity and philopatry in Greenshanks *(Tringa nebularia)* and Redshanks *(Tringa totanus). Proceedings of 19th International Ornithological Congress, Canada, 1986.*

TOMPKINS, S. C. 1986.
The theft of the hills: afforestation in Scotland. Ramblers' Association and World Wildlife Fund.

TYLER, S. J., & ORMEROD, S. J. 1985.
Aspects of the breeding biology of Dippers *Cinclus cinclus* in the southern catchment of the River Wye, Wales. *Bird Study, 33,* 164-169.

VÄISÄNEN, R. A., & RAUHALA, P. 1983.
Succession of land bird communities on large areas of peatland drained for forestry. *Annales Zoologici Fennici, 20,* 115-127.

VAN ECK, H., GOVERS, A., LEMAIRE, A., & SCHAMINEE, J. 1984.
Irish bogs. A case for planning. Nijmegen, Holland, Catholic University.

WALKER, G. 1985.
Inventories of ancient, long-established and semi-natural woodlands (provisional): Sutherland. Unpublished NCC report.

WALKER, G. 1986.
Inventories of ancient, long-established and semi-natural woodlands (provisional): Caithness. Unpublished NCC report.

WATSON, D. 1977.
The hen harrier. Berkhamsted, T. and A. D. Poyser.

WATSON, J., LANGSLOW, D. R., & RAE, S. R. 1987.
The impact of land-use changes on golden eagles in the Scottish Highlands. Peterborough, Nature Conservancy Council (CSD Research Report No. 720).

WEBB, A., REED, T. M., & WILLIAMS, T. D. 1983.
The Hebrides wader survey: did the observers record in the same way? *Wader Study Group Bulletin, 39,* 24-26.

WORLD WILDLIFE FUND-UK *et al.* 1983.
The Conservation and Development Programme for the UK. A response to the World Conservation Strategy. London, Kogan Page.

YEATES, G. K. 1948.
Bird haunts in northern Britain. London, Faber and Faber.

Appendix
Methods of ornithological surveys

Methods of data collection used in these surveys have been described in published reports (see section 3.1) and by Reed (1982b). For most moorland bird species, finding the nest of any given pair is too difficult for this to be a feasible means of population counting over the large areas which have to be surveyed. Observation of adult birds in their nesting territories thus has to be the principal basis of census. A territory mapping method was used, whereby a site was visited several times (usually four) in the course of the breeding season. A pair of observers walked a series of transect lines across the site 200 m apart; thus no part of the site was more than 100 m from an observer.

Once decided, the transect pattern was adhered to, with observers on subsequent visits using the same transects as on the first visit but reversing the direction of walking them on each occasion. Bearings were taken frequently to ensure that the same transect lines were used on each visit.

Weather can adversely affect the number and behaviour of birds seen. Recording was not attempted if there was strong wind (greater than Force 5 or even less in exposed areas), rain, low cloud or fog.

On each visit, observers recorded the birds seen from the transect lines. Each sighting was mapped, using a code (including details of behaviour) marked onto 1:10,000 maps in the field. Any double recording (i.e. when both observers saw and recorded the same bird during a transect) was corrected at the end of each line walked. At the end of each day a single composite map was produced. Not all bird species were recorded in the early years of the surveys. More recently, however, all species encountered have been recorded.

After each site visit the sightings were transferred to a summary map of the site for each species. At the end of the season it was thus possible to determine the number of territorial pairs of each species on each site from the clusters of sightings on the summary map. Birds were accepted as breeding or attempting to breed if one of the following was recorded (Reed, Langslow & Symonds 1983a):
□ a nest;
□ a pair with young;
□ a pair acting as if with young (cf Reed & Langslow 1985);
□ a bird or birds present in the same area on two or more occasions showing signs of attachment to the area.

Breeding densities are expressed in this report as pairs per square kilometre.

The analysis of dunlin sightings presents great problems. Dunlins breed semi-colonially, have small territories and do not move long distances to mob observers. Like snipe, they often behave cryptically. Problems of obtaining quantitative estimates of breeding numbers on machair habitat have been discussed by Reed & Fuller (1983), Reed, Williams & Webb (1983), Jackson & Percival (1983), Webb, Reed & Williams (1983) and Fuller, Green & Pienkowski (1983). Whilst dunlins breed at a lower density on blanket bog than on machair, many of the census problems are similar.

Greatest activity, and thus detectability, was found in the period 3-20 June. In estimating the number of pairs present, we followed the method used by Reed & Fuller (1983): the number of single birds 50 m or more from other birds seen during the period 3-20 June (or as close to that period as possible) was taken to represent a minimum number of pairs. This method of analysis differed from that employed for other species in being based on data only from the period of peak activity and not on clusters of records from the whole season.

After the breeding season, vegetation at each site was mapped and then divided into a grid of 200 m × 200 m squares. Details were recorded, using the provisional categories of the National Vegetation Classification (Birks & Ratcliffe 1980). As well as the vegetational composition of each square, details of the structure and physical features were recorded, such as presence of pools and dubh lochans, age and height of vegetation and the amount of regrowth after muirburn. The selectivity of breeding waders for particular areas within a site could thus be related to detailed habitat information (Reed & Langslow in press).

Seasonal timing of visits
In the course of five years' fieldwork, much effort was expended to identify biases in data collection and to evolve a standard methodology. Emphasis was placed on determination of optimal timing for census visits so that results would be repeatable and would accurately represent numbers of breeding waders (Langslow & Reed 1985; Reed in prep.).

It was found that ideally visits should be spaced about three weeks apart. Sites were usually visited a minimum of four times, with at least three visits between 1 May and 7 July. The most important periods were 16 May to 2 June, 3-20 June and 21 June to 7 July. These correspond to the main periods of territory establishment, incubation and fledging. At least one visit was made during each of the first two periods.

Diurnal timing of visits
In order to eliminate any biases due to differing detectability of waders at varying times of the day, diurnal variation in behaviour was investigated by Reed et al. (1983), Reed et al. (1985) and Reed & Langslow (in press). Dunlins, curlews, lapwings, snipe and golden plovers all showed significant diurnal differences in detectability.

Detectability was found to be highest in the early part of the day after dawn (before 09.00 hours), dropping to a low point in the early afternoon before rising to a second but lower peak of detectability in the early evening. This had important implications for the scale and timing of surveys.

Where wide-scale censusing was to take place, an early start would have resulted in a biased estimate for areas covered in the first few hours after dawn when compared with results obtained during the rest of the day (Reed et al. 1985). Avoidance of early morning starts provided compatible information from site to site. If large tracts needed to be completed within a single day, extension into the early evening activity peak did not seriously affect results (Reed et al. 1985). This was because the rise in activity and detectability in the evening was substantially less than in the early morning.

Validation of methods
The methods used for assessing dunlin densities were tested by Jackson & Percival (1983) during the survey of the breeding waders of the machair of the Outer Hebrides carried out by the Wader Study Group and NCC (Reed & Fuller 1983). The methods were checked against intensive nest searches and studies on colour-marked birds in several areas and were found to be consistent, although underestimating by about a third. Further checks were undertaken in 1985 and 1986 with similar results (Fuller 1985; Fuller & Percival 1986). As explained above, the same procedure for estimating the number of territorial pairs was used for the Upland Bird Survey.

The methods used to estimate densities of other territorial waders were tested in 1982 in an area of Wales where RSPB workers had independently searched for nests in an intensive survey of breeding waders. This enabled the method to be checked in terms both of overall numbers found by both methods and of the probability of nest location with respect to distance from the transect. Of seven golden plover nests found by RSPB surveyors, all were located by NCC transects, except that furthest (80 m) from the transect. As a proportion of nests found this was 86%.

In 1982 the transect method was tested at Kerloch, in Grampian Region, on an area of moorland intensively searched by the Institute of Terrestrial Ecology. The ITE personnel used trained dogs to locate all golden plovers in a moorland plot of about 500 ha. This was then surveyed by NCC workers, using the standard transect method. Of the 13 pairs of golden plovers known to be present by ITE, the NCC observers found 10 (77%) within the study plot. However, one of the pairs was known to have moved into a nearby field, where it was independently located by the NCC observers, who thus located 11 of the 13 territorial pairs (85%).

In no area did the territory mapping locate all nests. Estimates of breeding populations obtained this way must therefore be regarded as minima, particularly for dunlin.

Between-year comparisons of breeding wader populations

In order to investigate more closely between-year variations in wader breeding numbers and to see what implications these have for site assessments based on a single year's survey, Langslow & Reed (1985) surveyed a number of sites in Caithness and Sutherland in consecutive years. Overall, they found that between-year variation in the densities of moorland breeding waders was relatively low and showed no overall trend for all species. The spatial distribution of territorial birds on census plots in relation to the mosaic of habitat types did not vary significantly between years. This suggests that, although most Upland Bird Survey plots were surveyed in only one year, the results accurately represent the quality of the areas as habitat for breeding waders.